CONNECTING FAITH TO SIGHT

Study Genesis 1-11 from a Biblical, Scientific, & Historical Perspective

MELANIE NEWTON

JOYFUL
WALK
BIBLE
STUDIES

Connecting Faith to Sight: Study Genesis 1-11 from a Biblical, Scientific, & Historical Perspective

Published by Joyful Walk Press. Flower Mound, TX.

For questions about the use of this study guide or for bulk orders, please email us at melanienewton.com/contact.

ISBN: 979-8-9926517-1-3

Cover graphics created on canva.com, used by permission.

Scripture quotations unless otherwise noted are taken from the Holy Bible, New International Version ®, NIV ®. Copyright © 1973, 1978, 1984 by International Bible Society. Used by permission of Zondervan Publishing Company. All rights reserved.

The phrase "What we see in God's World agrees with what we read in God's Word" was adopted from the creation science teaching of Dr. Gary Parker.

Melanie Newton is the author of "Graceful Beginnings" books for anyone new to the Bible and "Joyful Walk Bible Studies" for established Christians. Her mission is to help women learn to study the Bible for themselves and to grow their Bible-teaching skills to lead others.

Joyful Walk Bible Studies are grace-based studies for women of all ages. Each study guide follows the inductive method of Bible study (observation, interpretation, application) in a warm and inviting format.

We pray that you and your group will find *Connecting Faith to Sight* a resource that God will use to strengthen you in your faith walk with Him.

Christ-Focused • Grace-Based • Bible-Rich

JOYFUL WALK PRESS
Flower Mound, TX

MELANIE NEWTON

Melanie Newton is a Louisiana girl who made the choice to follow Jesus while attending LSU. She and her husband Ron married and moved to Texas for him to attend Dallas Theological Seminary. They stayed in Texas where Ron led a wilderness camping ministry for troubled youth for many years. Ron now helps corporations with their challenging employees and is the author of the top-rated business book, *No Jerks on the Job*.

Melanie jumped into raising three Texas-born children and serving in ministry to women at her church. Through the years, the Lord has given her opportunity to do Bible teaching and to write grace-based Bible studies for women that are now available from her website (melanienewton.com) and on Bible.org. *Graceful Beginnings* books are for anyone new to the Bible. *Joyful Walk Bible Studies* are for maturing Christians.

Melanie is currently a disciplemaking trainer with Joyful Walk Ministries. She equips and encourages Christian women everywhere to pursue a lifestyle of disciplemaking. Her heart's desire is to encourage you to have a joyful relationship with Jesus Christ so you are willing to share that experience with others around you.

"Jesus took hold of me in 1972, and I've been on this great adventure ever since. My life is a gift of God, full of blessings in the midst of difficult challenges. The more I've learned and experienced God's absolutely amazing grace, the more I've discovered my faith walk to be a joyful one. I'm still seeking that joyful walk every day."

Melanie

OTHER BIBLE STUDIES BY MELANIE NEWTON

Graceful Beginnings Series books for anyone new to the Bible:

A Fresh Start (basics for new Christians)

Painting the Portrait of Jesus (I Am's in the Gospel of John)

The God You Can Know (the character of God)

Grace Overflowing (an overview of Paul's 13 letters)

The Walk from Fear to Faith (7 Old Testament women)

Satisfied by His Love (women who knew Jesus)

Seek the Treasure (study of Ephesians)

Pathways to a Joyful Walk (6 pathways to a joy-filled life)

HeartShine (selected Psalms)

Joyful Walk Bible Studies for growing Christians:

Adorn Yourself with Godliness (1 Timothy and Titus, also in Spanish)

Everyday Women, Ever Faithful God (Old Testament women, also in Spanish)

Connecting Faith to Sight (Genesis 1-11)

Graceful Living (the essentials for a grace-based Christian life)

Graceful Living Today (150 Bible-rich, Christ-focused devotions)

Healthy Living (Colossians and Philemon)

Heartbreak to Hope (the Gospel of Mark)

Identity: Sticking to Your Faith in a Pull-Apart World (Ezra thru Malachi)

Knowing Jesus, Knowing Joy (Philippians, also in Spanish)

Live Out His Love (New Testament women)

Perspective (1and 2 Thessalonians)

Profiles of Perseverance (Old Testament men, also in Spanish)

Radical Acts (Acts)

Reboot, Renew, Rejoice (1 and 2 Chronicles)

The God-Dependent Woman (2 Corinthians)

To Be Found Faithful (2 Timothy)

Resources for leading others

Be a Christ-Focused Small Group Leader

Leap into Lifestyle Disciplemaking

Painting the Picture of Jesus (the "I Am's" of Jesus lessons for children)

Teaching Children the God They Can Know (the character of God for children)

Download our catalogue and get resources for your spiritual growth at melanienewton.com.

Contents

Using This Study Guide.. 1

LESSONS

Lesson 1: "In the Beginning, God"...5

Lesson 2: The Six Days of Creation 15

Lesson 3: Humans, Home, and Family...27

Lesson 4: The Problem of Evil... 37

Lesson 5: The Lost World... 47

Lesson 6: The Flood and the Fossils... 57

Lesson 7: The New World & the Great Dispersion 73

RESOURCES

The Rest of the Story... 89

Discernment Practice: Interpreting What You Read and Hear 91

Sources 95

Using This Study Guide

Connecting Faith to Sight is a 7-lesson study of Genesis 1-11 from a biblical, scientific, and historical perspective. A proper understanding of Genesis 1-11 is important because nearly every Biblical doctrine is based on that section of the Bible. The lessons include scientific and historical facts that support the biblical text to give you confidence that what is revealed in God's Word is seen in God's world around you. The first three days study the biblical text. The fourth day is a "Creation Answers" article that covers evidence we see in our world that agrees with what we read in the biblical text.

THE BASIC STUDY

Each lesson includes core questions covering the passage narrative. These core questions will take you through the process of inductive Bible study—observation, interpretation, and application. The Inductive process is the best way to study the Bible. The process is more easily understood in the context of answering these questions:

- What does the passage say? (Observation: what's actually there)

- What does it mean? (Interpretation: the author's intended meaning)

- How does this apply to me today? *(Application: making it personal)* Questions identified as *"**Connecting Faith to Sight**"* lead you to introspection and application of a specific truth to your life.

STUDY ENHANCEMENTS

Study Aids: To aid in proper interpretation and application of the study, 6 additional study aids are located where appropriate in the lesson:

- Historical Insights

- Scriptural Insights

- Scientific Insights

- From the Hebrew/Greek (definitions of Hebrew and Greek words)

- Focus on the Meaning

- Think About It (thoughtful reflection)

Discernment Practice (optional): At the end of the book are short articles from the news containing both facts and assumptions. You can learn to discern anything you read or hear to separate the actual facts being presented from biased interpretations or assumptions being stated as fact.

For more information (optional): At the end of the lessons are suggestions for further research of subjects to gain information that would significantly enhance the basic study for you. Easy access to online resources is helpful, but two websites are especially recommended at the end of the lessons as quality sources for credible creation science information—Institute for Creation Research (**icr.org**) and Answers in Genesis (**answersingenesis.org**). Both are trustworthy sources of information and especially in their professional way of handling controversial subjects. Three other websites are especially beneficial for gaining historical, archeological, and theological information—**christiananswers.net** (click on the archeology information), **probe.org** (historical information and apologetics plus other topics of interest) and **bible.org** (a great source of biblical information).

OLD TESTAMENT SUMMARY

About 1700 years after God created everything, He sent judgment on a rebellious people through a worldwide Flood. He later separated the nations with different languages and scattered them from Babel. Abraham, Isaac, and Jacob were founding fathers of the Hebrew people. Sold into slavery, Joseph became a powerful foreign leader. The Israelites developed into a great nation for ~400 years in Egypt, until their deliverance from bondage. Then Moses took the people across the Red Sea and taught them God's Law at Mt. Sinai. Joshua led the Israelites into the Promised Land after a 40-year trek in the wilderness because of their unbelief.

During the transition toward monarchy, there were deliverer-rulers called "Judges," the last of whom was Samuel. The first three Hebrew kings—Saul, David, and Solomon—each ruled 40 years. Under Rehoboam, the Hebrew nation divided into northern and southern kingdoms, respectively called Israel and Judah. Prophets warned against worshipping the foreign god Baal. After the reign of 19 wicked kings in the north, Assyria conquered and scattered the northern kingdom. In the south, 20 kings ruled for ~350 years, until Babylon took the people into captivity for 70 years. Zerubbabel, Ezra, and Nehemiah led the Jews back into Jerusalem over a 100-year period. More than 400 "silent years" spanned the gap between Malachi and Matthew.

The 39 books in the Old Testament are divided into 4 main categories:

- "The LAW" (5 books)—the beginning of the nation of Israel as God's chosen people; God giving His Laws to the people that made them distinct from the rest of the nations.

- "HISTORY" (12 books)—narratives that reveal what happened from the time the people entered the Promised Land right after Moses died until 400 years before Christ was born.

- "POETRY and WISDOM" (5 books)—take place at the same time as the history books but are set apart because they are written as poems and have a lot of wise teaching in them.

- "PROPHETS" (17 books)—concurrent with the books of history and, except for Lamentations, reflect the name of the prophet through whom God spoke to Israel.

NEW TESTAMENT SUMMARY

The New Testament opens with the births of John (the baptizer) and Jesus. About 30 years after his birth, John challenged the Jews to indicate their repentance (turning from sin and toward God) by submitting to water baptism. Water baptism was a familiar Old Testament practice used when a Gentile converted to Judaism (to be washed clean of idolatry).

Jesus, God's incarnate Son, publicly showed the world what God is like and taught His perfect ways for 3 – 3½ years. After preparing 12 disciples to continue Christ's earthly work, He died voluntarily on a cross for humanity's sin, rose from the dead, and returned to heaven. The account of His earthly life is recorded in 4 books known as the Gospels (the biblical books of Matthew, Mark, Luke and John, each named after the compiler of the account).

After Jesus' return to heaven, the followers of Christ were then empowered by the Holy Spirit and spread God's salvation message among the Jews, a number of whom believed in Christ. The apostle Paul and others carried the good news to the Gentiles during 3 missionary journeys (much of this recorded in the book of Acts). Paul wrote 13 New Testament letters to churches and individuals (Romans through Philemon). The section in our Bible from Hebrews to Jude contains 8 additional letters penned by five men, including two apostles (Peter and John) and two of Jesus' half-brothers (James and Jude). The author of Hebrews is unknown. The apostle John also recorded Revelation, which summarizes God's final plan for the world. The Bible ends as it began—with a new, sinless creation.

DISCUSSION GROUP GUIDELINES

1. **Attend consistently** whether your lesson is done or not. You'll learn from the other women, and they want to get to know you.

2. **Set aside time** to work through the study questions. The goal of Bible study is to **get to know Jesus**. He will change your life.

3. **Share your insights** from your personal study time. As you spend time in the Bible, Jesus will teach you truth through His Spirit inside you.

4. **Respect each other's insights**. Listen thoughtfully. Encourage each other as you interact. Refrain from dominating the discussion if you have a tendency to be talkative. ☺

5. **Celebrate our unity** in Christ. Avoid bringing up controversial subjects such as politics, divisive issues, and denominational differences.

6. **Maintain confidentiality.** Remember that anything shared during the group time is not to leave the **group** (unless permission is granted by the one sharing).

7. **Pray for one another** as sisters in Christ.

8. **Get to know the women** in your group. Please do not use your small group members for solicitation purposes for home businesses.

Enjoy your Joyful Walk Bible Study!

Lesson 1: "In the Beginning, God"

DAY ONE STUDY

Creation is a hot topic. In fact, the validity of everything recorded in Genesis 1-11 is a hot topic, though not a new topic. The concept of some type of naturalistic explanation for the origin of the earth, universe, and its inhabitants has been around since the ancient Greeks. But it did not affect mainstream life until a couple of scientists promoted their ideas through books in the 1800's—1) Lyell who proposed that the earth was much older than what the Bible said it was and 2) Darwin who promoted biological evolution over long ages as an alternative to creation.

Our foundation of truth

Many Christians like studying the lives of Abraham, Jacob, and Joseph, but they avoid Genesis 1-11 because of controversy. Maybe you feel that way.

But God made sure Genesis 1-11 was recorded for us so we can know what He did and what happened to it. Those 11 chapters lay the foundation for the rest of the Bible. What is my point? You have to **know where to start**. Decide what will be your source of truth.

Did you know that nearly every Biblical doctrine is based on Genesis 1-11? Consider the doctrine of marriage, why humans have dominion over the earth, why there's death in the world, why we wear clothes, and most importantly, the reason Jesus came to die on a cross. All of these are dependent on a literal Genesis. The only reason we understand the meaning of sin is because we read about the actual rebellion of the first humans when they disobeyed God.

In Genesis 1-11, God demonstrates His sovereignty over all creation including humanity. If those chapters were taken out of the Bible, you must throw the rest away, especially Jesus and the cross, because it would make no sense. You have to know where to start. Understanding life as we know it starts with God's perfect Creation of everything in our world.

Knowing what happened leads to right conclusions

Genesis as the first book in the Bible is an account of the origins and history of all life and the universe. It tells of the creation of different kinds of animals and plants as well as the creation of Adam and Eve, the first people. Genesis explains the entrance of sin and death into God's Creation, and a judgment by water when a worldwide flood destroyed the original world. We are told of the event of the "Tower of Babel" that is important in understanding cultures and nations existing today. And in Genesis, we are given glimpses of God's plan for restoration of His creation.

It is impossible to study Creation, the Curse, the Flood, or the Tower of Babel without studying the relevant scientific and historical information as well. **Knowing** what happened in the past can give us a basis for explaining the present. If we start with what has been **revealed** about the events of the past, we have the ability to come to some right conclusions about geology, culture, geography, biology, archeology, and more. If we have the **revealed Word of One who knows everything and was there**, then a basis exists for having right interpretations about the facts of the present. If we ignore that, then we can never know for sure whether our interpretations are right!

Science is constantly giving us new information. Since science belongs to God, true science will always confirm God's revealed Word, which we know remains unchanged. We can have confidence that the Bible is true even in what is written in Genesis 1-11.

Taking captive every thought

Three scriptures are foundations for our trust in the Bible to be true:

*"**All scripture is given by inspiration of God**, and is profitable for doctrine, for reproof, for correction, for instruction in righteousness." (2 Timothy 3:16)*

*" … when you received the Word of God which you heard from us, **you accepted it not as the word of men, but as it actually is, the word of God**, which is at work in you who believe." (1 Thessalonians 2:13)*

*"Jesus said, 'Sanctify them through your truth: **your Word is truth**.'" (John 17:17)*

We start here. Knowing where to start helps us do what is written in 2 Corinthians:

*"We demolish arguments and every pretension that sets itself up against the knowledge of God, and we **take captive every thought** to make it obedient to Christ." (2 Corinthians 10:5)*

Consider the false messages you receive through the media and education about earth history and biological origins: Science is the source of authority and truth. Evolution is scientific fact and untouchable. The Bible is only a human work. Christians, stay out of anything that deals with origins. Stick to religion.

So how do you respond? Is compromise the only comfortable solution? Sadly, many religious leaders have done just that and are even holding science as the authority over biblical interpretation. This is most often because of two things: 1) the fear of being ridiculed and 2) not knowing that true science has facts that support Scripture.

Half of us are probably not affected by this. The rest of us are split personalities. Just say, "believe Genesis," and our thoughts are going crazy on one side of our brains! Our minds fill with "yeah, buts." And we think of every TV show, book, and magazine that teaches with confidence that this earth is billions of years old and humans evolved from apes. What do we do with those thoughts?

In this study, we want to help you take those thoughts captive so you can be whole again.

What if you are not convinced?

Please be assured that you can still be a Christian and believe that the earth is 4.5 billion years old and that you got here through evolution. Salvation is based upon faith in Jesus Christ alone.

We ask you to come to this study with an open mind. Give the biblical text a chance to give answers to your questions about origins. You will not get all your questions answered, maybe not even half. But **you can ask the One who knows all** to reveal to you some ways for you to believe that His record of beginnings is literally true. We will do our best to provide information to help you learn to think about this issue based on facts you can know—biblical, scientific, and historical. True science belongs to God. No believer needs to be afraid of it. True science will also provide information that agrees with what is revealed by God in the Bible.

Our approach for this study

The plain reading of the biblical text leads us to consider the earth and universe to have been created in six days in a perfect state, altered by the curse and death *after* creation, then completely reshaped by a global Flood—all less than 10,000 years ago. Here is a statement that we hope you will see to be very true as you do this study:

What we read in God's Word agrees with what we see in God's World. (Gary Parker, *Creation: Facts of Life*)

We hope this helps you connect your faith to what you see.

1. ***Connecting Faith to Sight:*** Reflect and respond to what you just read. What do you hope to learn through this study? What questions do you hope to have answered?

Let's get started. The best place to start is Genesis 1:1. Anyone who believes this verse will have little difficulty believing all the rest of what is written in the Bible.

Ask the Lord Jesus to teach you through His Word. Tell Him you are listening.

Read Genesis 1:1.

2. What does Genesis 1:1 tell you about God and the reality (or truth) of the world around us?

There are literally hundreds of verses about God as Creator. We will only study a few in this lesson.

3. Read Psalm 90:2. What does this verse tell you about God and our world?

4. Read Psalm 100. What does this psalm tell you … ?

    - About God—

    - About yourself—

    Think About It: When I come to the realization that God is God, and I am not, this is the beginning of true worship. I no longer limit God to my own understanding of Him and His works. Is this true for you as well?

5. The book of Job is full of discourses where Job and his friends try to figure out "why" Job is suffering and ultimately question God's actions. Read God's response to them all in Job 38:1-14. What does He say to Job and to us?

6. What do the following verses say about God as Creator?

    - Romans 1:18-21—

- Acts 14:15-17—

- Acts 17:22-31—

- Revelation 4:8-11—

Focus on the Meaning: The word genesis in Hebrew means "origin," and the Book of Genesis gives the only true and reliable account of the origin of all the basic entities of the universe and of life. Consider how Genesis contains the origin of (1) the universe and our solar system, (2) order and complexity, (3) the atmosphere and hydrosphere, (4) all life including humanity, (6) marriage, (7) language, (8) evil, (9) government, (10) nations and culture, (11) religion, and (13) the Jews. We will be covering all of these in this study.

Respond to the Lord about what He's shown you today.

DAY TWO STUDY

Ask the Lord Jesus to teach you through His Word. Tell Him you are listening.

Read Genesis 1:1.

From the Hebrew: The Hebrew word used for God in this verse is *Elohim*. *El* signifies the strong God. And what less than almighty strength could bring all things out of nothing? *Elohim* is plural. However, the verb "created" is singular, giving hint of the work of the Trinity—God the Father, God the Son, God the Holy Spirit.

7. What part did each take in Creation?

- Genesis 1:2—

- Genesis 1:26—

- John 1:1-5, 14-18—

- Colossians 1:15-20—

- Hebrews 1:1-2—

8. How does Jesus validate the authenticity of Genesis 1 in the following verses?
    - Mark 10:2-9—

    - Mark 13:19—

    Scriptural Insight: The Biblical account of creation is ridiculed by atheists, patronized by liberals, and often allegorized even by conservatives. The fact is, however, that it is God's own account of creation, **corroborated by Jesus Christ** (Mark 10:6-8), who was there! We are well advised to take it seriously and literally, for God is able to say what He means, and will someday hold us accountable for believing what He says! (Dr. Henry M. Morris, "The Logic of Biblical Creation," accessed at icr.org)

9. Read Job 40:1-2. Notice that God stops speaking and waits for Job's answer. Read Job's answers in the following verses. To what conclusions does Job finally reach about God?
    - Job 40:3-5—

    - Job 42:1-6—

10. **Connecting Faith to Sight:** After reading all these passages concerning God as Creator (Father, Son, and Holy Spirit), what conclusions do you reach about God?

Respond to the Lord about what He's shown you today.

DAY THREE STUDY

Ask the Lord Jesus to teach you through His Word. Tell Him you are listening.

Read Genesis 1:1-2:3.

11. In the Bible, any phrase repeated often is repeated for emphasis. It is important and valid. What repeated words or phrases do you observe? [We will cover the individual days in Lesson 2.]

12. What can you learn about God from His spoken word ("And God said, 'Let...'") and His sovereign power ("And it was so")?

13. According to Genesis 1:1-2:3, how many days were involved in God's work of creation?

Scientific Insight: Some claim a great span of time existed between Genesis 1:1 and 1:2, allowing for an earlier time when a populated earth was destroyed because of Satan's rebellion and then was restored to the way it appears now. This "Gap" is supposed to explain an old age for the earth plus layers of rocks and fossils but without evolution. The problem with this position is that any catastrophe that left the earth shapeless and empty would have destroyed everything as orderly as the continuous rock layers. So the Gap Theory is no longer considered a credible alternative to young earth creation.

14. God repeated the phrase "after/according to their kinds" ten times in Genesis 1. Through that phrase, what is God trying to tell us in terms that we can fully understand and not get wrong?

Read Exodus 20:11.

This is one of the ten commandments spoken by God to ~2 million Israelites at Mt. Sinai (Exodus 20). It was later spoken by God to Moses and written by the finger of God on stone tablets given to Moses for the people (Exodus 31:17-18). After those tablets were destroyed, God dictated the same words to Moses who wrote it down (Exodus 34:28). God declares He created everything in six days.

15. Using words from Genesis chapter 1, how does God define "day" so there should be no confusion?

From the Hebrew: The Length of a Genesis 1 Day—What does the Hebrew word for "day" mean in the context of Genesis 1 and in other contexts?

1) **Yom** (the Hebrew word translated "day") *always* means an ordinary day when used with a modifying number (1, 2, 3, 1st, 2nd, 3rd, etc.) all 358 times elsewhere in the Old Testament. Why would Genesis 1 be an exception?

2) The defining phrase "evening and morning" is used 38 times outside Genesis 1 and always means an ordinary day. Why would Genesis 1 be the exception?

3) Exodus 20:11 tells us that God worked for 6 days and rested for 1 day to give us a pattern to follow. Making 1 day = millions of years makes no logical sense!

4) **Yom** is never used to mean a definite, long period of time, such as the geological ages.

5) **Yamin** (plural, "days") is used 700+ times in the OT, always referring to a regular day.

6) God Himself distinguishes days from years in Genesis 1:14.

Why does it matter? We should not allow fallible humans to dictate what the words in the Bible mean. Take God at His Word. If it is obvious from the words and their context what the meaning is,. Then, accept this and not try to interpret God's Word on the basis of other's opinions. Instead, look for evidence supporting what God said.

16. ***Connecting Faith to Sight:*** How aware are you of the created world in your everyday life? What in creation fills you the most with a sense of God's glory?

Respond to the Lord about what He's shown you today.

DAY FOUR STUDY

Creation Answers—Science and the Age of the Earth

In her book, *A Woman's Journey to the Heart of God*, Cynthia Heald wrote, "Years ago, I discovered the freedom and power of making **pre-decisions** in order to take captive every thought and make it obedient to Christ." There are several pre-decisions that are the foundation for taking captive every thought we read, hear, and see to make it obedient to Christ regarding creation science.

Principle #1: In the beginning, God.

God existed forever. God created everything. This world is real, not an illusion. It had to come from somewhere. We have only 2 choices: 1) Nothing suddenly became something without a cause **or** 2) God, who always existed, created something out of nothing. There is no other choice. It is easier to believe that God always existed and then created something, than it is to accept that nothing created something. That's where we'll start.

Principle #2: God told us what and how He created.

When it comes to this physical universe, God told us in audible and written form that He created the heavens and the earth and everything in them in **six days** (Genesis 1; Exodus 20:11; 31:17-18; and 34:28. We covered that in this lesson. Yet many say Genesis 1 is a myth.

Let's take captive our thoughts here. God created humans with intelligence and the capacity for a trusting relationship with Him. Why would our almighty yet personal creator God start out His entire revelation with a story that isn't true if? Why would a God who cannot lie do that? To think so is illogical. So let's not tell God that He was wrong and didn't mean what He said.

God addressed that very presumption in Job 38 by asking Job, "Were you there when I laid the foundations for the earth?" (Job 38:4) What's the answer? Of course not. Who was there? God. Wouldn't He know what He did? The Bible is true and the source of truth. Human ideas are not.

Principle #3: True science is repeatable, testable, and falsifiable.

As defined by the Supreme Court, there is a difference between operational science and historical science.

- *Operational science* is repeatable, testable, and falsifiable. It describes natural laws and how they work in our world such as used in electronics, medicine, cars, etc. It describes what you can observe and repeatedly test in the present.

- *Historical science* is not repeatable, testable, and falsifiable. It describes observations made in the present in order to make "educated guesses" about the past. It is heavily influenced by one's personal belief system.

The two have become fused together into one term called "science." That's what really gets a stranglehold on us. A lot of us don't know the difference. But …

Principle #4: Science belongs to God.

He is the one who designed the DNA of every creature, how minerals and crystals form as well as what makes a bird fly and the sun produce light and heat. You can go to Him and ask for enlightenment on any issue. And you can trust that there are scientific answers out there that support what He said He did to create, curse, and flood this earth. Specific answers may not come today, but they may be there tomorrow if you are looking. Scientists are constantly studying this planet finding out new things. So …

Principle #5: Examine the evidence and ask questions.

Evaluate everything you read and hear in light of the Scriptures. It's remarkable what you find when you are willing to examine the evidence and look at facts. You can ask all the questions you want because God has left us enough clues to know that His Word, even in Genesis, is true.

No one can prove creation or evolution. But you can start by looking at what the Bible says then asking what would we expect to find if it is true? What do we find? We can examine facts in light of that model. Both evolutionists and creationists can look at the same facts but apply them to different models based on their personal belief systems.

Starting with the pre-decision that what God said is true in His Word, let's look for evidence in His world that the earth and universe may be younger than we have been taught.

Evidence for a Young Earth

One of the biggest hot buttons in the creation discussion is the age of the earth and universe. Where do we get the idea that the earth is billions of years old? It is not from the Bible, but the world's teaching uses human theories to reinterpret God's Word.

Is it really radical to believe in a young age for the Earth?

A warning light goes off in your head. How can I believe in a young age for the earth? Haven't scientists proven that the earth is very old—billions of years? What about all the evidence that the earth and universe are billions of years old? It's just too radical to believe in a young earth.

This is the answer: What evidence? There is more verifiable evidence that the earth is less than 10,000 years of age than that it is very old. Did you know that?

Why is there such an insistence in the science world and media that the universe and earth are billions of years old? Think about it. If you eliminate God and any divine aspect to creation, then you depend strictly on naturalistic processes and lots of time! Evolution, if true, needs a long time for it to take place. So the evolutionary scientist looks for any way to "prove" lots of time has passed.

What's the prevailing theory?

About 15 billion years ago (BYA) compressed matter exploded with a Big Bang. As this explosion took place, the elements in the periodic table cooked their way into existence and were hurled outward. They began to condense and to cool, galaxies were formed as well as nebulas. Out of one of these spiral nebulas, a solar system condensed and in this particular solar system, the planet Earth spun out of a molten fireball about 4.6 BYA.

What questions should you ask?

How did that compressed matter suddenly appear out of nothing? We've already looked at what God said He did and how. The biblical model is easier to believe—that God created everything out of nothing, mature, and fully functioning from the beginning.

Did you know that there are about 100 "process" dating methods recognized? Nearly all of them (95%) suggest much younger ages for the earth and life than billions of years (earth's magnetic field decay, supernova remnants, amount of mud on the sea floor, helium decay measurements and more). The popular radioisotope dating methods that supposedly give very old ages for rocks are not reliable. When using radioisotope dating on rocks of known age (volcanic rock that humans know the exact year it was formed), the results will show millions of years rather than anywhere close to the actual date. Why should anyone accept the old ages touted by some scientists as reliable? You don't need to do so.

And neither rocks nor fossils have birthdates stamped on them when they are collected. Fossils are usually dated by the false assumption of evolution, which demands old age. We'll cover the evidence from fossils in a later lesson.

One of best evidences that the earth is a lot younger than thought comes from diamonds. Diamonds form as carbon atoms are subjected to intense pressures deep in the earth and brought to the surface through volcanic activity. According to secular scientists, they should be billions of years old and have no measurable carbon-14 in them. But they do! Carbon-14 has a short lifespan (no more than 50,000 years). The presence of carbon-14 suggests that the earth could be no more than ~50,000 years old.

If you look at the available scientific evidence that is reliable, you can see that most process clocks agree that the earth is considerably less than that. So holding to a young age for the earth based on the biblical text is completely logical and scientifically verifiable.

Four facts to remember when it comes to the age of the earth:

- The earth has not been scientifically **proven** to be billions of years old.

- The Bible teaches a literal **6-day** creation.

- **Many scientist**s do believe in a literal 6-day creation about 6,000 years ago.

- There are many scientific evidences for a **young earth**, including new evidence regarding nuclear decay.

What we read in God's Word agrees with what we see in God's World.

For more information

There are several sources of very credible scientific information that support Genesis 1-11 as revealed to us by God. Regarding what we covered in this lesson, go to the Institute for Creation Research at **icr.or**g or Answers in Genesis at **answersingenesis.org**. Do a search on topics such as age of the earth, radioisotope dating, distant starlight, creation days, young earth evidence, etc.

You can also sign up for the free *Acts and Facts* magazine at **icr.org**. This wonderful bimonthly publication presents creation science research in easy-to-understand terms. There is even a page just for kids.

Discerning facts from assumptions

What are the facts in a news report? What are the assumptions? You have to separate the two to get to the truth.

So much information that is proclaimed in the media (news, magazines, movies, internet, and books) is biased against the biblical truth presented in the Bible. Often, there is a morsel of fact mixed with a ton of assumptions that are presented as facts, even if they are pure speculation. Every believer needs to be able to discern the truth from speculation.

To practice discerning facts from assumption, go to the "Discernment Practice: Interpreting What You Read and Hear" page in the back of this book. Learn how to follow the process given.

Lesson 2: The Six Days of Creation

DAY ONE STUDY

Understanding terminology

Christians today are ensnared in the trap of not thinking critically. In the Bible, we read that God created all things in 6 days. But, we've heard that all educated people know that evolution has been proven. We need to train ourselves to think, to recognize the difference between scientific facts that can be observed in the **present** and ideas about the **past** that are used to interpret those facts we see in the **present**.

There are four prevailing theories regarding the origin of the universe, the earth, and life itself predominant in our culture—both secular and Judeo/Christian. Here is a brief description of each.

- **Naturalism (Naturalistic Evolution)**—Naturalism allows no place for God, miracles or divine intervention. Natural processes can explain everything. Evolution is based on naturalism.

- **Theistic Evolution**—God created matter, and then everything else evolved with or without occasional help from God. Some say God directed the evolutionary process. It's basically evolution with God added. This position denies the truth of Scripture.

- **Old Earth Creation or Progressive Creation**—The "days" of creation were long periods of time or ages during which God created the universe and all its contents. Evolution and the geologic column basically fit with the sequence of events of the creation week although you have to twist and creatively interpret the text to make it do so. Many Christians have bought into this teaching because it sounds plausible and avoids being ridiculed. However, in Mark10:6, Jesus combined Genesis 1 creation days with the Genesis 2 creation of man and woman as happening at the same time, not millions of years later.

- **Young Earth Creation or Literal Creation**—Young earth creationists adhere to the plain reading of the text. This position believes that God did His creating just as He told us in Genesis 1 and 2. This position is the easiest to describe and explain but the hardest for many to accept even when plainly understood.

You might ask, "What do I do? How am I supposed to decide what to believe?" Remember that we have the revealed Word of One who knows everything and who was there! Trust His Word, study it, consider what you would expect to find if it is true, and then examine the evidence in the world.

Definitions to know for this study:

- "Creation" is defined simply as *the work of God in bringing all things into existence*. Only God is eternal—everything else in the universe had a beginning. True creation is creation *ex nihilo* (out of nothing) and is not merely a reworking of materials already in existence.

- "Evolution" is defined, in its broadest sense, as t*he theory that all things have been derived by gradual modification through natural processes from previous materials*. According to this concept, all forms of life have developed from earlier, simpler forms, and even life itself spontaneously came into existence through a complex organization of previously nonliving chemical molecules.

When God first began His creating, the earth was described as "formless and empty" (v. 2). "Formless" just means it needed shape. It was covered with water (the deep). "Empty" means unfilled. God spent 2½ days *separating and gathering* to give the earth form (1:3-10). God spent 3½ days *making and filling* to remove the emptiness (1:11-28). Let's see what He did each day.

Ask the Lord Jesus to teach you through His Word. Tell Him you are listening.

Read Genesis 1:3-5 (the first day).

1. The fact of the day/night cycle beginning in these verses is evidence for regular rotation of the earth. This does not contradict science but is mutually supported by scientific observation. God does not tell us the source of light on the first day since the sun, moon, and stars weren't created yet. Light is given off by other physical sources such as radioactivity, crystals, volcanoes, rocks moving against one another, fire, gaseous clouds, etc.

   • Is it really significant that we know the exact source of the light?

   • Read Revelation 22:5. What will be our future source of light?

   • How does the above verse help you understand that light can exist apart from the sun (Genesis 1:3-5)?

   Think About It: God created light, so this light is not Himself. He divided the light from the darkness. Also He refers to day and night and evening and morning. Because God is eternal, this also indicates the creation of time for the benefit of His creation. We are not told the source of the light if other than God Himself. Based upon what God said He did on Day 1, we would expect to find evidence of day and night being continual and universal. This is what we find.

Read Genesis 1:6-8 (the second day).

2. God spent the first 3 days separating and gathering to give the earth form. What was separated and gathered on Day 2?

3. God gives us more information about the expanse (sometimes translated as "firmament" or "vault") in Genesis 1:9, 20.

   • What information is given about this expanse?

   • So how might the earth have looked at the end of Day 2?

Scientific Insight: No one really knows what the "waters above" means. It could possibly refer to water vapor in our atmosphere like that above tropical areas today—not necessarily thick but more worldwide than at present. If so, we might expect to find evidence of a worldwide tropical climate, lush vegetation, plus greater size and longevity of creatures due to shielding from incoming radiation by the water expanse, whatever that was. We do find tropical plants and animals all over the globe in every rock layer, at the poles and on top of the highest mountains, and giant animals and plants. And we find evidence of exceptionally long human lifespans.

4. *Connecting Faith to Sight:* Think about the amazing nature and provision of water. Dwell on its unique properties AND why these are so important to life. What do you like best about water? Where do you enjoy it the most?

Respond to the Lord about what He's shown you today.

DAY TWO STUDY

Ask the Lord Jesus to teach you through His Word. Tell Him you are listening.

Read Genesis 1:9-13 (the third day).

5. Through separating and gathering again, how does God give more form to the earth during the first part of Day 3?

Scientific Insight: As a result of the raising of crustal blocks to form land out of water, the surface waters that had been covering them would have drained rapidly off the land. So, massive erosion must have occurred. We see evidence of this at the base of the Grand Canyon where there are thick folded and metamorphosed sedimentary layers with no plant of animal fossils in them. It is likely these were all deposited catastrophically during the erosion and deposition of Day Three of Creation Week. (Dr. Andrew Snelling, "Thirty Miles of Dirt in a Day," *Answers* Magazine, Vol. 3 No. 4, p. 29)

6. How might the Earth have looked midway through Day 3?

Scientific Insight: From the text, it appears that the land was all in one land mass. This doesn't preclude having rivers and lakes on the land, though, as described in Genesis 2. You would expect evidence of one large landmass with set boundaries between the ocean and the land. Scientists have found evidence of a supercontinent that has since split apart. Continental shelves and slopes are set boundaries between the ocean and the land. God spoke the supercontinent of the pre-Flood world into existence on Day 3 of Creation Week.

7. After forming the earth and space, God began filling it to take away its emptiness. Describe what He did next (Genesis 1:11-13).

8. Give some examples of the types of plants described here.

Scientific Insight: According to evolution, seed plants didn't "evolve" until after bacteria, invertebrates, fish, amphibians and some land animals. Once again, the Days of Genesis cannot be matched up with the theoretical "Geological Ages." Based on what God said He did on Day 3, we would expect to find evidence of fully-formed and functioning plants, fruit, and seeds from the start of life on earth. We find just that. Pollen is found in some of the oldest rocks supposedly millions of years older than the first pollen-forming plants. Pollen comes from flowering plants that produce fruit with seeds in them.

9. Why were plants created? See also Genesis 1:29-30 and 2:9a.

Scientific Insight: Plants do not generally grow on bare rock. When God said, "Let the earth sprout vegetation," He had already prepared whatever was necessary for plant life to grow. Yet, the generation of soil from rock supposedly takes many years. If you could go back in time to this point on the third day of creation, the earth would appear to be very old. This appearance of age is required for a mature creation to be fully functioning from the beginning. Yet, it definitely skews "age-dating" methods that rely on observation of present processes and the rate at which soil is now being produced.

10. ***Connecting Faith to Sight:*** Ponder the marvelous diversity and complexity of plant life in general as well as the many functions of plants in serving us—food, shade, and enjoyment. What are your favorite plants and why? Meditate on the Creator's loving hands fashioning these for you.

Read Genesis 1:14-19 (the fourth day).

11. With what does God fill the sky?

> **Think About It:** God separated light and darkness. Light filled the expanse that day. These heavenly bodies were fully functioning in a mature state from the beginning. Actually, the creation of light before the sun is evidence of authenticity of Genesis 1. Only an eyewitness would dare to write that. Also, plants can live without the sun for 24 hours or more I cover my plants in cold, icy weather for several days. But they could not live without the sun for millions of years. Another reason to believe the Genesis account to be true.

12. How and why did God create the lights in the sky? See what these verses say about it.

- Genesis 1:14-18—

- Psalm 19:1-6—

- Jeremiah 31:35—

13. Since much of pagan worship included worship of the sun and other celestial bodies, what message was God giving by waiting until Day 4 to create the sun, moon and stars? See also Deuteronomy 4:19; 17:2-5.

> **Scientific Insight:** Did you know there is no scientific proof that the sun is billions of years old? "There is no evidence based solely on solar observation that the sun is 4.5-5 billion years old. I suspect that the sun is 4.5 billion years old. However, given some new and unexpected results to the contrary, and some time for frantic recalculation and theoretical readjustment, I suspect that we could live with Bishop Ussher's value for the age of the earth and sun. I don't think we have much in the way of observational evidence in astronomy to conflict with that." (Dr. John A. Eddy, Solar Astronomer at the High Altitude Observatory at Boulder, CO)

Bishop Ussher (~1600 AD) calculated the earth/sun to be ~6500 years old based on Gen. 1-11. Today's scholars using the Masoretic text of the Old Testament plus the determined date of 2166 B.C. for Abram's birth according to the Jewish calendar can say that Earth's "birthday" was around 4128 B.C.

14. **Connecting Faith to Sight:** On a dark night, spend some time outside and watch the stars while reading Psalm 104 by flashlight. If you like to study constellations, also read Amos 5:8. Or watch a sunrise or sunset. Feel free to use any creative means to express what you see and your response God's creation (poem, song, drawing, craft, essay).

DAY THREE STUDY

Ask the Lord Jesus to teach you through His Word. Tell Him you are listening.

Read Genesis 1:20-23 (the fifth day) and Isaiah 45:18.

15. God formed the earth to be inhabited. All the materials necessary for animal life—water, air, light, plants, and the basic materials of the earth—were now available. With what living creatures did God fill the earth on Day 5?

Scriptural Insight: Are plants alive? The Bible makes a sharp distinction between plants and animals. On Day Three, God commanded the inanimate earth to "*bring forth*" plants while on Day Five, He "created...*every living* creature that moved." At this point, and on Day Six to follow, He instituted the concept of giving "life" (Hebrew *nephesh*) to nonliving matter—something He did not do for plants. The Bible never refers to plants as living. They grow, flourish, wither, and fade but they do not "live" or "die." Plants do not have *nephesh*, nor "breath of life" nor "blood" (Leviticus 17:11). They are biologically alive but not biblically "living creatures."

16. Name some types of water creatures represented by the descriptions given here.

From the Hebrew: It is possible that these "sea monsters (NAS), great whales (KJV), sea creatures (NIV)" actually included the great marine dinosaurs. The word in the original Hebrew (*tanniyn*) is actually the same word translated "dragons" in later Scriptures. Remember that the term "dinosaur" wasn't even invented until the mid-1800's.

17. The Hebrew term translated "winged bird" denotes anything that flies. Name some types of creatures represented by the descriptions given.

18. Current evolutionary theories say reptiles (land animals) evolved into birds. Notice the order of creation in vv. 20-21. What is from the Genesis account conflicts with that theory?

Scriptural Insight: The creation order contradicts the proposed evolution of the universe and life on the planet at more than twenty points. For example: Marine creatures weren't created until Day 5, but evolution says that they came before the land plants. Seed plants didn't "evolve" until after bacteria, invertebrates, fish, amphibians and some land animals, but creation says they came first. Bees "evolved" before flowers. You have to twist Genesis 1 to make it match up to evolution.

Read Genesis 1:24-2:1 (the sixth day).

19. With what did God fill the earth on Day 6?

20. Name the types of land creatures represented by the descriptions given.

The grand climax of God's creative activity are humans (vv. 26-31). We will study this in Lesson 3.

21. What instructions were given regarding food (vv. 29-30)? To which creatures? What are the biological and ecological implications of this?

22. **Connecting Faith to Sight:** Spend some time thinking about various animals. Read Job 38:39-39:30 as God tells about the specialness of some He created. What are your favorite animals and why?

23. In v. 31 and 2:1, how does God evaluate His creation?

24. Why is it not logical that disorder, decay, suffering, or death could have occurred or could be occurring in God's world as described in Genesis 1?

God did not do a halfway job. Everything mentioned thus far was fully formed and functioning, each ready to do its preordained job. We are not going to get all our questions answered although most do have logical answers already supplied through Scripture.

> **Think About It:** God told us how long creation took and how complete it was. God rested because He was finished. Nothing formless or empty remained. The biblical model reveals a fully functioning, mature creation: (1) Continents with topsoil and rivers flowing from a source of water, (2) Rocks with crystalline minerals in them—some already beginning the radioactive process to add warmth and other elements to our planet, (3) Stars visible from earth, (4) Plants bearing seed and fruit; bees pollinating flowers, (5) Marine animals swimming; birds flying, and (6) Land animals creeping and running; adult Adam and Eve talking—all capable of reproducing. Creation is complete.

Read Genesis 2:2-3 (the seventh day).

25. What occurred on the 7th day and why?

> **Think About It:** God rested and gave a command to the Israelites that they should dedicate a day of rest from their labors (Exodus 20:8-10). Could it be that God created humans with a need for rest and relaxation from work? Have you stopped to rest for at least part of a day this past week? How?

26. ***Connecting Faith to Sight:*** God created a well-ordered, functioning world for humans to enjoy.

- Read Hebrews 11:1-6. What is the role of faith in considering God's creation?

- Read Romans 1:20-25. What should be our response to God's self-revelation in the created world?

- What are the consequences for refusing to recognize God as Creator?

Respond to the Lord about what He's shown you today.

DAY FOUR STUDY

Creation Answers—Designed, Not Descended, Part 1

What evidence from life in the present would we expect to find if evolution were true?

First, you would expect to see life coming from non-life. This is the biggest problem for evolutionists. No one has ever produced life from non-life in a laboratory. It's not occurring spontaneously anywhere on the planet either.

Then, you would need to find beneficial mutations that continually create new systems in every living creature observable in the present. These mutations would be producing billions of transitional forms—a hypothetical organism that is "between" evolutionary stages. We should have transitional structures for every part of every system of every living creature in the present. We don't see this at all.

And we should see things getting better—no death due to aging or disease. Any process that could generate the evolution from amoebas to humans would have overcome death due to disease and aging. This is not what we actually see in our world.

What evidence from life in the present would we expect to find if creation is true as Genesis 1 says?

You would expect to see:

- Evidence of intelligent design in all living things.

- Separate, distinct kinds of creatures with much variation within them.

- Death and disease because of the Curse

- Extinction of many creatures because of the enormous ravaging work of the Flood. We'll cover this in a later lesson.

This is precisely what we do see in our world. Let's look at evidence for the first three more closely.

Intellligent design in all living things is obvious.

Every time students look through a microscope, telescope, or at the *real world*, they are bombarded with evidence for creation. Even the simplest single-celled organism is unthinkably complex. A virus, which isn't even complete in itself, can baffle our best scientists and all our technology!

All of life is governed by the marvelously complex genetic code, which contains not only design and order, but what is equivalent to written information. This DNA code must not only be written correctly, but the rest of the cell must read it and follow its instructions to live and reproduce. This code had to be present at the origin of life. How could it have written itself, and how could all the various parts of the cell learn how to read and obey it? Impossible!

Separate, distinct kinds are evident; transitional forms are non-existent.

For amoeba-to-human evolution to be possible there has to be a mechanism that adds **new genetic information** to the DNA recipe of cells—information that performs a new function that was not previously present. No such mechanism is known. The proposed mechanism is through a series of mutations that add new information to the DNA. Upward progress is supposedly made because of natural selection and beneficial mutations that give an organism the best chance of surviving a given habitat.

Mutations are inherited copying errors in the DNA "recipe" of an organism. Evolutionary theory requires some mutations to go "uphill"—to **add new information**. The mutations we observe are either neutral (information not effectively changed) or go "downhill"—the parent genetic information is lost or scrambled. In rare cases, a mutation may benefit a creature in a particular environment. It looks beneficial, but in every case these involve loss of information. For example, wingless beetles cannot produce beetles with wings again. It's all about information.

If evolution were true, we should find billions of transitional forms representing the progression from one creature to another in the present. Besides mutations, the examples commonly cited as "evolution happening today" usually involve variation or adaptation coupled with natural selection. Variation and adaptation are evidences of design more than of evolution.

- **Variation:** God placed within the DNA of creatures the potential for variety in certain parts of the creature. Everyone gets 2 sets of genes—1 from mom, 1 from dad. So, lots of potential for variation in hair color, eye color, skin color. But, there are limits on natural genetic variability. Humans will always have hair rather than fur and two eyes rather than 10. Noses can change shapes, but gills are never substituted for noses. Limited variety is present within every type of creature. There is variation in dogs, but all of them are still dogs; variation in cats, but they are all cats; variation in humans, but we are still all humans—and so on.

- **Adaptation:** Adaptation is a design feature in a plant or animal that makes it more suited to live in its environment by using pre-existing information to make the necessary adaptations to an environment. The best explanation for variation and adaptation is the new biological theory called "Continuous Environmental Tracking" based upon engineering principles. Organisms continuously trace environmental changes. A creature's self-adjusting innate mechanisms produce solutions to changing conditions. For example, when people change their elevation from sea level to mountaintop or warm climate to cold climate, their bodies will produce more red blood cells to fuel their "engines" with oxygen and warmth. The ability to adapt must be there before it can be selected! It's all about information.

Death and disease are universal because of the Curse.

All creatures get sick, get old, and die. Any evolutionary process that goes from amoeba to human would have solved this problem ages ago. It's illogical to think otherwise. The bottom-line is this: evolution is not being observed in the present.

God created all plants, animals, and humans according to kinds with lots of room for variation and adaptation within kinds. Each was subsequently affected by the curse. **THERE IS A CREATOR!**

What we read in God's Word agrees with what we see in God's World.

For more information

For those who like research: Consider the astronomical basis of the day, month, and year in contrast to the lack of such basis for the week. Why is this important? Did any cultures in ancient history besides the Jews have a 7-day week?

Regarding what we covered in this lesson, go to the Institute for Creation Research at **icr.org** or Answers in Genesis at **answersingenesis.org**. Do a search on topics such as distant starlight, evidence of special creation rather than evolution, the created kinds, continuous environmental tracking, and more.

To practice discerning facts from assumption, go to the "Discernment Practice: Interpreting What You Read and Hear" page in the back of this book. Learn how to follow the process given.

Lesson 3: Humans, Home, and Family

DAY ONE STUDY

Ask the Lord Jesus to teach you through His Word. Tell Him you are listening.

Read Genesis 1:26-31.

1. What information is given about God's creation of humans in vv. 26-27? Notice the use of the plural pronouns for God both here and in Genesis 3:22.

2. Image and likeness are synonymous in both the Old Testament and the New Testament. What might it mean to be made in the "image of God?" Clues are in the following verses. Try to find other verses to support your answer.

   • Genesis 1:26-31—

   • Genesis 2:15-17, 19-20—

   • Colossians 3:9-14—

   Think About It: We are to be walking, talking, visible representatives of the invisible God.

3. How many times does God use the word "created" in Gen. 1:27? _____ What significant information is He trying to get across to us?

4. Together, as male and female, God gave them a function on earth that was equal in responsibility and accountability. What are their responsibilities?

5. So what does God declare in verse 31 about His creation that includes both man and woman?

6. ***Connecting Faith to Sight:*** God created man and woman with intellect, emotion, and a will through which every human can choose to love and serve God. We are not puppets nor are we programmed to do the same things year after year as the animals do. Contemplate this "endowment" from the Creator.

   • In what ways are you thankful for this ability to choose?

   • How does knowing the truth that you are created in God's image affect how you feel about yourself?

   • Read Psalm 8. How is God mindful of your life (v. 4)?

Respond to the Lord about what He's shown you today.

DAY TWO STUDY

Ask the Lord Jesus to teach you through His Word. Tell Him you are listening.

Genesis 2 is supplemental, not contradictory to what we read in chapter 1. Jesus quoted from both chapters together in Mark 10:6 showing that He considered it to be a unit. Genesis 2 reveals important details about God's highest creation—humans, the one creature intended to have a spiritual relationship with God.

> **Historical Insight:** It is likely that as Moses wrote the book of Genesis, he used records that had been handed down through the generations, each record associated with the man who wrote it. Genesis 2:4a says, "This is the *account* of the heavens and the earth when they were created." There are 9 more of these lines found in Genesis seemingly at the end of a section, like a signature at the end of a letter. The first one is God's, obviously given by direct revelation. The second one is Adam's in Genesis 5:1 and is called a *book or written account.* In fact, the word "generations" (Hebrew *tholedoth*) was translated into English as

"Genesis." It also appears in Matthew 1:1 *"The book of the generation (genesis or genealogical history) of Jesus Christ."*

It is now known that writing was common before Abraham, so it is reasonable to think these early records were also written. By the way, the many intimate details and descriptions throughout Genesis indicate that these were eyewitness accounts probably written on stone or clay tablets. Moses, guided by the Holy Spirit, then compiled and organized them into a continuous narrative. God used the same method in inspiring the other historical books of the Bible which were either written by eyewitnesses or from the direct verbal or written testimonies of eyewitnesses (for example, 2 Chronicles 9:29; 12:15).

Read Genesis 2:4-17.

7. Describe the Garden of Eden as God prepared it for Adam and Eve. What is growing there? [Note: shrub/plant of the field are those requiring cultivation.] What is the temperature? What about the watering system?

Historical Insight: We don't know where the Garden of Eden was located. The Flood would have totally restructured the surface of the Earth, wiping away all traces of the Garden except what may be found as fossils. God placed an angel at the entrance to the Garden to keep pre-Flood humans from returning. That was no longer necessary once the Flood obliterated it. After the Flood, Noah and his descendants gave familiar names to the new rivers and places they encountered as they migrated, just as American settlers have done for 400 years.

8. The phrase "breath of life (lives)" is an interesting one. Read Job 12:10 and Job 34:14-15. What further information is given to us?

Focus on the Meaning: On the previous days of creation God said, "Let there be ... " But the creation of humans is different. Genesis 2:7 says, "The Lord God *formed* (means to shape and mold with loving care) the man ... and *breathed* into his nostrils the breath of life (literally, *lives*)." The name used for God is YHWH, the personal God, and He gave Adam "the breath of life." Animals also had the breath of life in them, but God personally gave it to Adam as humanity's representative.

9. *Connecting Faith to Sight:* Considering the "Focus on the Meaning" above,

    • What does this tell you about God's relationship to humans as opposed to the animals and the rest of creation?

- What does knowing this truth mean to you personally?

10. Read the command God gave to Adam in Genesis 2:16. Adam already had knowledge of "good." All he had seen and experienced was "good." Discuss God's command regarding the Tree of the Knowledge of Good and Evil—what it involved, what would be gained/lost, and why choosing to eat from that tree would be so bad.

Focus on the Meaning: This verse reveals that God gave humans the emotion of fear before sin occurred. Fear is a normal human emotion designed by God to alert us to danger so that we will take action against it. Adam was told there was a danger in eating the fruit from that one tree. The action to take was not to eat it.

11. The Tree of Life is mentioned in other passages as well. Glean information given about it in …

- Genesis 3:22-24—

- Revelation 2:7—

- Revelation 22:2,14,19—

Think About It: What was God's intention by placing the Tree of Life in the Garden of Eden? Although we don't know its exact composition, we do know that it represented a choice. True love involves a choice. There has to be the possibility of rejection in order for there to be true love.

12. By having access to the fruit of the Tree of Life, God showed that His will and intention for them was life—continual physical life. The results of eating from the Tree of Knowledge of Good and Evil was death, best translated "dying, you shall die." Why do you think the penalty was so severe for choosing to disobey?

13. ***Connecting Faith to Sight:*** God prepared a beautiful garden for the man and woman, one in which they would both work and enjoy. Spend some time in a garden setting this week—garden, yard, or park—where you can enjoy the trees, various shades of color in leaves and flowers, fragrances, etc. A garden can be an inspiring place. Ponder the beauty of God's creation. God made all those wonderful plants for you! How does a beautiful garden inspire you?

Respond to the Lord about what He's shown you today.

DAY THREE STUDY

Ask the Lord Jesus to teach you through His Word. Tell Him you are listening.

Read Gen. 2:4-25.

14. God created humans with the tools of intelligence, observation, and language. Discuss the task God gave to Adam in Genesis 2:18-20.

15. As God brought certain animals to Adam, what is soon revealed?

> **From the Hebrew:** Although made in the image of God, a perfect handiwork, Adam was not complete. In fact, this is the first thing in creation that God calls "not good." God decided that Adam needed a helper. The Hebrew word *ezer* means "one who assists another in reaching complete fulfillment." It is used nineteen times in the Old Testament: four times as equal to equal (Psalm 89:19) and fifteen times as God to humans (Psalm 10:14; Isaiah 41:10). It is NEVER used as an inferior helping a superior.

16. Compare the timing and method of the creation of Adam to that of Eve.

- What are the similarities?

- What are the differences?

> **Think About It:** Could all the events recorded on the 6th day of creation really have taken place during a normal, 24-hour-type day? The Bible does not say that Adam named **all** the living creatures on that day, only those God brought to him from three groups—livestock (cattle), beasts, and birds. In the opinion of many researchers, the purpose of Adam's task may have been to discover his unique aloneness. The number of animals would be determined by how long it would take Adam to fully get the point!! So would the God who spoke everything into existence have difficulty creating the animals, creating Adam, planting a Garden, talking to Adam, bringing certain animals to Adam, waiting for Adam to name the animals and creating Eve—all in a day's time? The answer is a definite NO!!

17. Discuss Adam's response to Eve. [Note: The Hebrew word translated "woman" (*ishah*) is the feminine form of *ish*, which is translated "man."]

> **Think About It:** She is his (man's) peer in capacities of intellect, moral worth and sensibilities. She can think, feel, imagine, reason; she can sell goods, plan buildings, make statues, diagnose diseases, construct philosophies, or write epics. In a word, what is open to a man as a human being is open to her. (Dr. Allen Ross, Hebrew Scholar)

18. ***Connecting Faith to Sight:*** We must understand who we really are as women, by God's design, in order to have a proper view of our value as persons and our responsibility to function as God intended. God did not made neither man nor woman to be totally independent of one another. Nor did He make woman to be inferior to man—intellectually, emotionally, morally, or spiritually. Knowing these truths will free you to become the woman God wants you to be. Spend some time thinking about God creating you as a woman rather than as a man. List all the benefits and blessings of womanhood. Use your creativity to praise your Creator for His marvelous design.

19. Read Genesis 2:21-25 again. Notice the progression of "leave," "cleave" and "unite." Define these three words using a dictionary:

- Leave—

- Cleave—

- Unite—

20. What was God's intention for Adam and Eve and their descendants regarding future marriage?

- Malachi 2:15-16—

- Jesus' support of this in Matthew 19:4-6—

21. The first marriage was performed ("husband" and "wife" are used in Genesis 3). This was God's institution. He invented it to be one man, one woman for life. Discuss Genesis 2:25. What does being naked without feeling any shame mean?

> **Think About It:** Marriage is the decision to identify with another person and adopt a **'team approach to life.'** Identification is the building of a **partnership**. We must therefore 'die to singleness in order to be alive to marriage.' (Tim Stevenson)

22. Connecting Faith to Sight: Read Ecclesiastes 4:9-12.

- *If you are married*—God has given you the gift of a marriage partner for intimate community so that you will not be doing life "alone." Consider how to give your **marriage priority** this week over other relationships and activities—over children and their activities, church responsibilities, school, job, house, perfectionism, shopping, etc.

- *If you are single*—How do you connect with other adults in intimate community? Consider what you might do to make sure you are not doing life "alone."

Respond to the Lord about what He's shown you today.

DAY FOUR STUDY

Creation Answers—Designed, not Descended, Part 2

There are billions of dead things buried in rock layers all over the earth represented by bones, shells, molds, leaves, teeth, skin imprints, footprints, dung. Scientists have been examining the fossils and the rocks in which they are found for hundreds of years.

As a way of reconstructing evolution's past history, scientists have set up a "Geologic Time Scale" where the rocks with the "most *primitive*" fossils are assigned the oldest ages, and the rocks with the "most *advanced*" fossils are considered the youngest with various stages in between. However, there is no place on earth where all the rock layers, with the kinds of fossils listed in the geologic time scale, are found in the sequence listed. The geologic time scale is a nice graphic, but it doesn't represent reality.

What evidence from life in the past would we expect to find if evolution were true?

We would expect to see that fossils show a progressive change from one organism to another—the transitional forms I mentioned in the last "Creation Answers" discussion at the end of Lesson 2.

What evidence from life in the past would we expect to find if creation is true as Genesis 1 says?

We would expect to see that each basic kind appears in the fossil record complete, with no incomplete ancestors. We would also expect to see fossils similar to living forms, possessing some variations. That does represent reality. This is what we find in the fossil record. Each creature on earth is designed, not descended.

Complex and diverse life appears abruptly in the fossil record.

So many kinds of organisms suddenly appear at the lowest level that it's called the "Cambrian Explosion." Thousands of very complex organisms suddenly appear in these rocks such as fish, sea urchins, and trilobites that have a segmented body and complex vision. No ancestral forms can be found for any of these. They appear in the fossil record fully formed and distinct.

Evolutionists cannot explain how this happened even if they dig deeper into "older" layers to hopefully find the "earlier stages of developing forms." So the evidence says, **"It's designed, not descended."**

Each basic body style has been present right from the start.

New body plans have not appeared since the lowest levels of fossils. Shells look like shells; eyes look like eyes. For instance, clams are found in the bottom layer, the top layer, and every layer in between. There are many different varieties of clams, but clams are in every layer and are still alive today. There's no evolution, just clams!

The existence of many living fossils also challenges the supposed hundreds of millions of years of earth history. A living fossil is one that is the same in fossil and living form. Some of these fossils are missing from intervening strata that supposedly represent millions of years of evolutionary time, again indicating that there were no time gaps. For example, supposedly "530 million-year-old" starfish, jellyfish, and snails look very much like those living today.

Transitional forms are absent.

Darwin predicted that the fossil record would show numerous transitional fossils. Yet, millions of fossils later, all we have are a few disputed ones. Popular magazines frequently publish articles claiming the latest transitional form has been found, having a nice artistic drawing on the front only to have to retract their statements in a short sentence tucked away in the back somewhere a few months later. Beware of what you see in Natural History Museums. Don't assume the skeletons are real fossils. Often, complete skeletons are reconstructed from 1-2 bones. Then, flesh is put on that skeleton based upon the artist's evolutionary bias. As you read descriptions and news releases, remember to separate the facts from the assumptions.

Did you know that there are thousands of examples of "Out of Place" fossils? That means they aren't where they are supposed to be according to the Geologic Time Scale. So don't think every fossil that is found fits neatly into the expected time category. You just don't hear about the ones that don't.

Human evolution is imaginary.

Nobody has demonstrated step-by-step how any ape-like set of features could have mutated into any of the long list of specifically human traits, including skeletal arrangements that enable a human's peculiar and efficient "knees pointed forward" manner of walking. The Creator of all life said in His Word that He specially created humanity in His image, and the fossil evidence bears this out.

Proposed fossil examples of human ancestors are soon proven to be "just one more dead end in the questionable human evolution parade" as one researcher said. Some fossils are not informative enough to accurately identify. Thus, any given "hominin" fossil is either unknown, or from a human (Neandertal), or from an ape (Lucy).

Neandertal and Cro-Magnon people are now considered to be people groups living at the same time as other human people groups elsewhere, having reached Europe first during the worst part of the Ice Age. They intermarried with each other as many fossils show such a mix of traits. Both groups were fully human. *Just as expected!*

Based upon the actual fossil evidence, the conclusion is this: all life, including human life, is **designed, not descended** through an evolutionary process.

What we read in God's Word agrees with what we see in God's World.

For more information

Regarding what we covered in this lesson, go to the Institute for Creation Research at icr.org or Answers in Genesis at answersingenesis.org. Do a search on topics such as Garden of Eden, tree of life, human family, Neandertals, marriage in history, creation of woman, differences between males and females, or similar phrase. Select an article to read or video to watch.

> To practice discerning facts from assumption, go to the "Discernment Practice: Interpreting What You Read and Hear" page in the back of this book. Learn how to follow the process given.

Lesson 4: The Problem of Evil

DAY ONE STUDY

Ask the Lord Jesus to teach you through His Word. Tell Him you are listening.

Read Genesis 3:1-6.

1. In Hebrew, the word *Ha-satan* (literally, "the accuser") describes Satan's nature as an accuser, adversary, and hater. [Note: By the time of 1 Chronicles, the article was dropped and *Satan* had become a proper name.] What is revealed about this creature in the verses below?

    *   Job 1:7, 9; 2:2—

    *   John 8:44—

    *   1 Peter 5:8—

    Scriptural Insight: Who is this serpent? The serpent, one of the subtlest of God's creation, is recorded to have spoken to Eve. The Apostle Paul accepted the historical account of the serpent—and its ability to deceive—as a real phenomenon. To the Christians in Corinth, the apostle wrote, "I am afraid that as the *serpent* deceived Eve by his cunning, your thoughts will be led astray from a sincere and pure devotion to Christ." (2 Corinthians 11:3). In Revelation, John describes Satan when he notes, "And the great dragon was thrown down, that ancient *serpent*, who is called the Devil and Satan, the deceiver of the whole world (Revelation 12:9; 20:2)."

2. In Genesis 3:4-5, notice how the serpent rephrased what God said to Adam in Genesis 2:16-17 about the Tree of the Knowledge of Good and Evil. Which portion of the serpent's statements are true and which are false?

    *   True—

    *   False—

3. Compare Eve's response in Genesis 3:2-3 with what God actually said and did in 2:9, 16-17. So in what way did Satan lead Eve … ?

    *   To doubt God's Word—

- To distrust God's character—

Read Genesis 3:7-13.

We will now see the effects of sin on us as individuals, on our relationship with others, and our relationship with God.

4. Once Adam and Eve had eaten the fruit, *"the eyes of both of them were opened, and they realized they were naked"* (v. 7, NIV). Compare this with Genesis 2:25 and what the serpent "promised" in 3:5.

    - What had actually been gained?

    - What had really been lost?

      Scriptural Insight: What is evil? Evil is not something God created as a counterpart to good (as often taught in other religions). In the beginning, God created all things "very good" (Genesis 1:31). Evil has only a parasitic existence. It is a diminishment, misuse, or perversion of something good that God created. Why does God allow evil to exist? We don't know the answer to that question. But we do know that the possibility of evil is inherent in God's choice to create beings with free will. God cannot simply annihilate evil without destroying free will, but He can and will defeat it. In the future, God plans to destroy evil forever (Revelation 20-22). We also know that God limits evil. Satan has boundaries beyond which he cannot cross (Job 1-2). God takes evil seriously and so should we. The presence of evil around us and the seeds of self-destruction in us should remind us of our need for dependency on Christ. Anything that causes us to rely on Him more is good for us. (adapted from *"When World-Views Collide,"* Tim Stevenson, Mark Series, Message #47.)

5. How did Adam and Eve try to solve their problem of "nakedness?"

6. Discuss what happened in vv. 8-11 including the apparent emotions of Adam and Eve.

7. ***Connecting Faith to Sight:*** Though Adam and Eve were hiding, God sought them. He knew where they were. Read Psalm 139. Is it really possible to hide from God? Does His consistent presence give you comfort? Why or why not?

8. Discuss the attempts of both Adam and Eve to avoid responsibility for their individual actions (vv. 12-13).

9. ***Connecting Faith to Sight:*** Do you have trouble accepting responsibility for your own actions? Who or what do you usually blame?

Respond to the Lord about what He's shown you today.

DAY TWO STUDY

Ask the Lord Jesus to teach you through His Word. Tell Him you are listening.

Read Genesis 3:14-19.

10. After the questioning, God proceeded to pronounce consequences on all the parties involved—the serpent (& Satan), the woman, and the man. Discuss what God said to each one and what it means. Look for evidences of God's grace to each as well.

- The Serpent (vv. 14-15)—

- The Woman (v. 16)—

- The Man (vv. 17-19)—

Scriptural Insight: Death entered the world. Instantly Adam and Eve died spiritually (separated relationship from God) until they were restored by God's grace (Genesis 3:21) their faith in Him (Genesis 4:1). And they began to die physically until they actually returned to dust. And physical death of living creatures, including man, began with God's judgment on man's sin. If you think about it, adding millions of years of animal death before the creation and Fall of man contradicts and destroys the Bible's teaching on death and nullifies the redemptive work of Christ on the cross. It also makes God into a bumbling, cruel creator who uses (or can't prevent) disease, natural disasters, and extinctions to mar His creative work without any moral cause but calls it all "very good." If God originally intended death to be an integral part of His creation, then God should allow death to continue into eternity. But He doesn't. All the prophecies of the future talk about a restoration to harmony in nature, animals eating only plants, and man living long on earth. Paul calls death the last enemy. Revelation 22:3 says the curse will be removed in the new Earth. Death will be vanquished for good.

Remember the literal translation of the Hebrew at the end of Genesis 2:17 is "dying, you shall die." Adam and Eve began to die physically on that very day, just as God promised. As we will soon see, their source of perpetual life was removed from their reach. But physical death was not the only aspect of life involved. We will study spiritual consequences later in this lesson.

11. Eve rightly recognized that she had been deceived (v. 10). Consult a dictionary for the definition of "deceive."

12. What can and should you do to root out deception and prevent it from happening to you?

- 1 Peter 5:6-10—

- Ephesians 6:10-13—

The good news is that we don't have to live in fear and defeat because of temptation and the problem of sin around us daily. As the one who remained faithful in temptation, He (Jesus) became the model for all believers when we are tempted. We have an Advocate!

13. Read the following verses. In times of temptation, what does the Bible say about the living presence of Jesus helping and encouraging us?

- Hebrews 4:15—

- 1 Corinthians 10:13—

- Hebrews 2:18—

14. Satan experienced victory over Adam and Eve in the garden. But God declared his ultimate defeat in Genesis 3:15. What did God say would happen?

Focus on the Meaning: The term *seed* has a biological connotation and usually refers to the man's part in conception. One could be a physical seed of woman, "her seed," only if there was no male seed involved in conception. This necessitates a virgin birth.

15. Genesis 3:15 refers to the *seed* (singular) of the woman—a woman's descendant. Who is the only man born to a woman without benefit of a man's sperm? Read the following verses:

- Luke 1:31-35—

- Galatians 4:4—

Scriptural Insight: God made His first Messianic promise in Genesis 3:15. He proclaimed essentially the first Gospel to Adam and Eve, promising victory over Satan and victory over his power against them.

16. Who or what is meant by "the seed of the serpent?"

- John 8:44—

- Acts 13:10—

- 1 John 3:8—

17. **Connecting Faith to Sight:** Sin has consequences. But as the hymn goes,

> Marvelous grace of our loving Lord, grace that exceeds our sin and our guilt.
> Yonder on Calvary's mount outpoured, there where the blood of the Lamb was spilt. Grace, grace, God's grace, grace will pardon and cleanse within.
> Grace, grace God's grace, Grace that is greater than all our sin. (*Grace Greater Than Our Sin*, Julia Harriette Johnston)

Read and reflect on Romans 5:12-21. How has God's plan benefited you?

Respond to the Lord about what He's shown you today.

DAY THREE STUDY

Ask the Lord Jesus to teach you through His Word. Tell Him you are listening.

Read Genesis 3:20-24.

18. Discuss the meaning of Eve's name in Genesis 3:20.

19. Compare Genesis 3:21 with Genesis 3:7. Why weren't the clothes Adam and Eve made good enough?

20. What was better about God's provision of clothing?

21. Watching God take the life of innocent animals to make clothes for Adam and Eve must have been a sobering experience for them. You may have wondered why God would choose the sacrifice of animals to pay the price for human sin, as it seems to be so cruel. As an object lesson, what could every animal sacrifice remind humans about their own sin?

> **Scriptural Insight:** The fig leaves of Adam and Eve's own invention were insufficient, for what was actually involved here was the need for spiritual covering, something which would be accomplished only by faith in shed blood (Hebrews 9:22). God graciously provided for their nakedness with "coats of skin." Perhaps they silently and sorrowfully waited as God selected two of their animal friends, probably two sheep, and shed the innocent blood before their eyes. They learned, in type, that an atonement (or covering) could only be provided by the shedding of blood upon the altar. (Dr. Henry Morris, *The Beginning of the World*, p. 77)

From Genesis 2:9 and 16, we learn that Adam and Eve were to eat from the Tree of Life. We can assume they did. It seems that life was to be perpetuated for man and woman as they consistently ate of the Tree of Life. In Genesis 3:22, God chose to withdraw the privilege of eating from this tree and live forever in a sinful state.

22. God prevented Adam and Eve from eating from the Tree of Life by driving them out of the garden. Why is this both a punishment and an act of mercy?

23. ***Deeper Discoveries (optional):*** What other information is given in Scripture about the cherubim and the sword? See Ezekiel 1:4-8; 10:1-22; 28:14; and Revelation 4:6-8.

24. What remained the same and what was different regarding Adam's job description before the Curse as opposed to afterward? Compare Genesis 1:28; 2:15; and 3:17-19,23.

- Before—

- After—

25. ***Connecting Faith to Sight:*** Almighty God loves man and woman—stubborn, disobedient, selfish, and frail—so much that He provided suitable clothes for them that provided protection and warmth. And He planned to redeem them and us by making it possible for us to once again have perfect fellowship with Him. Read John 1:29 and John 3:16 and meditate on the meaning.

Respond to the Lord about what He's shown you today.

DAY FOUR STUDY

Creation Answers—Cursed Creation

John Milton's epic poem *Paradise Lost*, the second longest poem in English literature begins with these words:

> Of Man's first disobedience, and the fruit
> Of that forbidden tree whose mortal taste
> Brought death into the World, and all our woe,
> With loss of Eden, till one greater Man
> Restore us, and regain the blissful seat,

Genesis 2-5 contain the account of "Paradise" and "Paradise Lost" for us.

Genesis 2—The perfect world

Genesis 2 reveals important details about God's highest creation—human kind, the one creature intended to have a spiritual relationship with God. From Genesis 1 and 2, we glean some specific details.

- Animals and humans were vegetarian so there was harmony in nature.

- God specially designed both the man and the woman with purpose.

- God made man and woman (not the animals) in His image and established a relationship with them.

- Humanity and most of the animal world would have continual life. Death was not expected.

- God pronounced His creation "very good" at the end of the 6th day and rested from creating.

Any evidence of disorder, suffering, decay, and death which we now see in the present world or in the records of the past cannot possibly be attributed to anything occurring during the six days of Creation. Something obviously corrupted this "very good" perfect world, turning it into the world we see today. The answer is found in 1 Corinthians 15:21 which says, *"By man came death."*

Genesis 3—Human sin and "Paradise Lost"

Corruption aptly describes what happened to God's perfect world. Corruption affected all of life. Do you know that many cultures have in their traditional histories (before ever having contact with a Jew or Christian) an account of a perfect world that was marred by disobedience? A "paradise lost." Some even involve a tree.

Death entered the world. Instantly Adam and Eve died spiritually meaning they had a separated relationship from God until they responded to Him by faith after God covered their sin. They also began to die physically until they physically returned to dust. Since Adam and Eve were appointed to exercise dominion over the earth, their dominion also would begin to "die."

Paul described this in Romans,

> *Therefore, just as sin entered the world through one man, and death through sin, and in this way death came to all men, because all sinned. (Romans 5:12)*

"World" is the Greek word "kosmos" which is used to refer to the universe and the entire earth. Death did not enter only the experience of humanity but also the whole creation, which would include the animal kingdom—the other living creatures besides humans.

A cursed creation affected everything! The ground became reluctant to yield its food. Instead it required human toil, sweat, and tears before anyone could eat. Some creatures and plants were changed. Women would experience pain and sorrow in childbearing. Some plants would now produce thorns and thistles rather than just useful fruit and vegetables.

Relationships were affected. Pain, decay, suffering and aging would become part of the daily experience, as would animals eating one another. We will see later that humans weren't told to eat meat until after the Flood (Genesis 9). The curse not only affected humanity, but also the earth and all of creation itself. Romans 8:20-22 describes how the whole creation is groaning because of sin, waiting for its restoration.

> *For the creation was subjected to frustration, not by its own choice, but by the will of the one who subjected it, in hope that the creation itself will be liberated from its bondage to decay and brought into the glorious freedom of the children of God. We know that the whole creation has been groaning as in the pains of childbirth right up to the present time. (Romans 8:20-22)*

Did God change creatures so they would become fierce, eat meat, and die? Perhaps the defense / attack structures were already present and used for different purposes before the Curse. Did God create them with those capabilities in the first place and just "turned it on" after the curse? Or God could have made design alterations after the Curse to allow such defense / attack structures. There's argument for both ideas.

Consider sharp teeth. Even today, creatures with sharp teeth do not always use them to rip other animals apart. The giant panda's sharp teeth are used for bamboo munching. The fruit bat eats fruit with its fangs. Some originally vegetarian animals become accustomed to eating meat when they are exposed to it. So it's not unlikely to think that they could have all been vegetarians to begin with, but God allowed them to "branch out."

What about disease? While some organisms cause disease in one type of creature, those same organisms are beneficial to other creatures. Many *E. coli* bacteria are very helpful to the human body, especially aiding digestion.

Physical changes were definitely made to plants producing thorns and to the serpent in losing its legs. These changes were passed along to their offspring. Humans (and animals) would no longer live perpetually but age and die.

Both of these ideas may be true. We just don't know how God did it. But He did something that permanently affected the perfect "very good" world. Through it, God is allowing humanity to experience what it wants—life without God.

What we read in God's Word agrees with what we see in God's World.

For more information

Regarding what we covered in this lesson, go to the Institute for Creation Research at **icr.org** or Answers in Genesis at **answersingenesis.org**. SEARCH a word or phrase from today's passage such as sin, angels, Satan, curse, or death and suffering. Select an article to read or a video to watch.

To practice discerning facts from assumption, go to the "Discernment Practice: Interpreting What You Read and Hear" page in the back of this book. Learn how to follow the process given.

Lesson 5: The Lost World

DAY ONE STUDY

Ask the Lord Jesus to teach you through His Word. Tell Him you are listening.

Though no longer perfect, life moves on. And God is in it.

Read Genesis 4:1-16.

> **From the Hebrew:** In this story, we find Cain and Abel providing an offering to God. The word used for offering in this context is *minchah*, which means "meal (cereal) offering; offering; tribute; gift; sacrifice." *Minchah* is often used to refer to any "offering" or "gift" made to God, whether it was a "vegetable offering" or a "blood sacrifice." In later Old Testament texts, *minchah* is commonly used to designate grain or cereal offerings, usually mixed with oil, to be offered alone or with a burnt animal sacrifice. The Hebrew word *olah* means a burnt animal sacrifice and first occurs in Genesis 8:20, when Noah built the altar to the Lord and sacrificed burnt offerings on it. The same Hebrew word for offering, *minchah*, was used for the offerings of BOTH Cain and Abel. Scripture does say that Abel brought "fat portions of some of the firstborn of his flock" which means he brought the "choicest parts." (*Vines Complete Expository Dictionary of Old and New Testament Words*)

1. What insight do the following verses give about the Lord's acceptance of Abel's offering but not Cain's?

   • Hebrews 11:4—

   • 1 John 3:10-15—

2. In the following references, why was the offering unacceptable?

   • 1 Samuel 15:3, 21-23—

   • Malachi 1:6-10—

3. In the following references, what is considered a pleasing sacrifice or better than a sacrifice?

   • 1 Samuel 15:22—

- Psalm 51:16-17—

- Proverbs 15:8—

4. So after answering the previous two questions, what does God really want?

5. How does Cain respond to the Lord looking with favor on Abel's offering but not his own (v. 5)?

6. God personally communicates with Cain. What does the Lord ask Cain (vv. 6-7)?

7. After explaining that Cain should take responsibility by doing what is right in order to be accepted, God also points out that if "you do not do what is right, sin is crouching at your door; it desires to have you, but you must master it." Explain what you think God means by this.

8. Cain did not master sin. What were the consequences—for himself, his brother, his family, and his descendants?

9. Throughout this whole passage, on whom or on what is Cain's focus?

10. **Connecting Faith to Sight:** Read Psalm 139:23-24. *Memorize* these verses and *pray* them often this week. Be sensitive to what the Holy Spirit reveals to you about your own heart. Think about what areas of your life need to be yielded to the Holy Spirit so that you will do what is right and have mastery over sin.

Respond to the Lord about what He's shown you today.

DAY TWO STUDY

Ask the Lord Jesus to teach you through His Word. Tell Him you are listening.

Read Genesis 4:10-17.

11. Discuss the following:

- God's judgment upon Cain—

- God's mercy to Cain—

> **From the Hebrew:** The Hebrew word translated "city" in verse 17 refers to any permanent settlement. It is logical to conclude that Adam and his clan were already building a permanent settlement in their homeland by this time.

12. Just as the location of Eden before the Flood is unknown, the location of Nod is also unknown (vv. 16-17). The name Nod just means "wandering, exile." Discuss the changes that then took place in Cain's life as a result of God's judgment upon him.

Historical Insight: The question about where Cain got his wife has fed scoffers for years. Scripture does not say he found a wife in Nod nor does it prevent him from having already been married before he murdered Abel. Since Adam and Eve had sons and daughters (Genesis 5:4), and since God had not prohibited marriage between close family members (not until the time of Moses), it is reasonable to conclude that Cain married a sister or a niece. According to the historian Josephus, Jewish tradition held that Adam and Eve had 56 sons and daughters. Considering their lifespan of > 900 years and their perfect physical condition, it is not impossible to believe. (*The Genesis Record*, Dr. Henry Morris, p. 142)

Read Genesis 4:18-24.

The remaining portion of Genesis 4 gives a picture of life in the antediluvian (pre-Flood) world. We have limited information about that first human civilization, later to be completely destroyed by the great Flood. Archeological excavations reveal only post-Flood cultures. Some cultures have semi-legendary recollections of the world's first "golden age." This brief record in Genesis chapter 4 is the only reliable account we have of that first time period.

13. Notice the type of heritage that Cain passed down to his descendants, particularly Lamech who is in the seventh generation from Cain. The word translated "killed" in verse 23 means "to slay, slaughter" and is the same word used of Cain's murder of Abel. Comment on Lamech's activities and attitudes towards others, himself and God.

14. What aspects of *civilization*, *culture* and *intellectual achievement* BEFORE THE FLOOD are found in the following verses from Genesis 4-5? Comment about ALL the skills involved in each. Be specific!

    • 4:17—

    • 4:20—

    • 4:21—

    • 4:22 [Think through the phrase "all kinds of tools."]—

    • 5:1 [The word translated "book" means "written account."]—

Scriptural Insight: Some commentaries suggest that the above evidences of culture and civilization are the by-products of a godless people, such as Cain's descendants, since the activities are described in Cain's lineage rather than in Seth's. The truth is that God has gifted humans with the abilities needed to do these things and even given directions to godly men in Scripture to perform similar works of "civilized" behavior (Exodus 25; 1 Kings 5-8; Deuteronomy 3:6-25; Matthew 25:14-28).

15. Are these the works of primitive, uncivilized, slow-witted "cave-men" humans? Explain.

Historical Insight: The record of Cain's descendants reveal that metal tools of all kinds were available to use, and musical instruments stimulated the emotional and aesthetic senses for them as it does for us. Some created poetry. Some lived as nomads, engaging in the domestication and herding of livestock. And there was writing as we have already discussed. Lamech's two wives were noted for their fashion sense: Ada means "gorgeously adorned" and Zillah means "one whose presence is announced by the tinkling of her jewelry." Although subject to death, humans used the intelligence God gave them. These are not the works of primitive, uncivilized, slow-witted "cave-humans." Job who lived shortly after the Flood also knew about metalworking. Sadly, much of archeology is influenced by the concept of evolution.

Read Genesis 4:25-26; 5:1-5.

16. After the murder of Abel, God provided another son for Adam and Eve, and another line of descendants began. Compare 5:1 with 5:3. What could be the meaning of Adam's statement about having a son in his likeness?

Focus on the Meaning: To 'call' on God's name (Genesis 4:26) is to summon His aid. *Calling* in this sense constitutes a prayer prompted by recognized need and directed to One who is able and willing to respond. Often this Hebrew verb represents sustained communication as when God called to Adam in Genesis 3:9. It may also mean "to proclaim" or "to announce." This is not the beginning of prayer, because communication between God and humanity existed since the Garden of Eden. Nor is it an indication of the beginning of formal worship, since formal worship began at least as early as the offerings of Cain and Abel. (*Vines Complete Expository Dictionary of Old and New Testament Words*)

Read Genesis 5:6-32.

So, what is the point of including these selective genealogies in Genesis? Perhaps these are to show the contrast in heritage between *men who focus on pleasing themselves* AS OPPOSED TO *men who focus on pleasing God*.

17. What is known about Seth's descendant Enoch? Read the following verses to derive your answer.

   • Genesis 5:18-23—

   • Hebrews 11:5-6—

   • Jude 14-15 (a sample of Enoch's preaching)—

Historical Insight: Assuming there are no "gaps" in these chronological genealogies, and that the years are not anything other than normal years (the Jewish year was 360 days), there seems to have been 1656 years from the creation of Adam to the Flood. The record is perfectly straightforward and is obviously intended to give both the necessary genealogical data to denote the promised lineage and also the only reliable chronological framework we have for the antediluvian (pre-Flood) period of history. Taking the recorded ages at face value, it is interesting to note that Adam lived until Lamech, the father of Noah, was fifty-six years old. Most likely the oldest of the living patriarchs maintained the primary responsibility for preserving and promulgating God's Word to his contemporaries.

The scriptures indicate that the antediluvians (those who lived before the Flood) lived to very great ages (averaging 912 years!) and that most families were quite large. The antediluvian world was substantially different from our own ... not much time had passed since the perfectly created human body began to be plagued by such mutations that became hereditary. As we study Genesis further, it will be noticed that life spans begin a slow and steady decline after the Flood, showing evidently that they were connected with the pre-Flood environment. There is no reason not to take this list in Genesis 5 as sober history. The names are repeated in I Chronicles 1:1-4 and Luke 3:36-38. This confirms that they were accepted as historical by the later Biblical writers of both Old and New Testaments. (Adapted from *The Genesis Record*, Dr. Henry Morris, pp. 153-155)

[The long lifespans of Genesis 5 are discussed in "Creation Answers—The Lost World and Its People" at the end of this lesson.]

18. **Connecting Faith to Sight:** If you have a relative who prayed or is praying for you and shared the gospel with you, spend some time thanking God for that person.

Respond to the Lord about what He's shown you today.

Day Three Study

Ask the Lord Jesus to teach you through His Word. Tell Him you are listening.

Read Genesis 6:1-13.

It is speculated that by this time (~1656 years after Creation), the population of the earth would have been in the millions. Longevity of life and large families would have contributed to a large population. Genesis 6:1 bears witness to the fact that humanity had indeed multiplied.

Who were those sons of God (verses 1-2)?

The phrase "sons of God" has caused interpreters difficulty for years! No one knows with confidence who they were. They might have been demon-possessed humans who participated in the daily activities of life and promoting rebellion against God, even exhibiting superhuman strength as they did in Jesus' day. Or they could represent believers in God marrying unbelievers and joining in the rebellion against God. What we do know is that over time, their children became excessively wicked. We are told God's observation and judgment against the humans living at the time. So let us concentrate on the rest of this passage in the questions rather than speculating on this topic.

Regardless of interpretation, Genesis 6 describes how corrupt the world had become as a result of sin in less than 1700 years.

19. Who were specifically identified in Genesis 6:4?

From the Hebrew: *Nephilim* comes from a Hebrew word meaning "to fall" so the word can be translated "fallen ones." However, it can also refer to those who fall upon others such as attackers/bullies. These were the same as "the men of renown." The Greeks translated the Hebrew word *nephilim* into *gigantes* (giants). The term came to be used after the Flood of men who were excessively tall (giants) and/or fierce, behaving as bullies. See Numbers 13:31-33. Their exploits of strength and violence made them famous in song and fable in all nations in the ages following the Flood. Perhaps these were the source of the heroes of Greek, Roman, and Sumerian mythology. To men of later times, especially rebellious men, they were revered as great heroes. *Gigantes* also applies to the physical characteristics of pre-Flood animals. Nearly all modern animals and plants have larger fossil ancestors. Giant cockroaches; armadillos 6 feet tall; birds with wing spans of 52 feet; and, of course, dinosaurs—20 feet tall, 50 feet long and weighing 7 tons! It would be reasonable to conclude there were "giant" humans in those days.

20. From God's perspective, what were these people like?

- Verse 5—

- Verses 11-13—

21. What was God's response to what He saw (v. 6)?

22. What does He plan to do about it and when (vv. 3, 7)?

23. Whom does God notice to be the exception to the corruption on the earth (vv. 8-9)?

24. **Connecting Faith to Sight:** When all around us is total depravity, we individually have the choice of God's ways or the world's ways. How does it make you feel to know that God notices even just one person, such as He did Noah, and such as He does you?

Respond to the Lord about what He's shown you today.

DAY FOUR STUDY

The Lost World and Its People

What could have been the population of the earth ~1600 years after Creation? Take the average lifespan of 912 years plus 9 children per family. Just including only those families listed in Genesis 5 and taking into consideration the violence prevalent on the earth, there may have been at least 2 million people alive at the time of Noah. If there were that many humans alive, we might expect to find lots of human fossils, right? Yet, most human fossils are found in what is classified as post-Flood sediments or in caves, not in Flood deposits. Why is that? Good question.

Since the waters of the Flood took weeks to cover the earth completely, humans could have climbed to higher ground, grabbed onto floating debris, then died and decomposed rather than being buried. The speed at which water and other forces eliminated all traces of 43,000 victims in the 2004 Asian tsunami absolutely shocked scientists. Dead humans bloat and float rather than sink.

The Big Three—characteristics of the pre-flood Earth

What was that "Lost World" like? Three particular evidences scream out about this lost world. I call them the Big Three.

1) Long life spans for humans

2) Appearance of a worldwide tropical "greenhouse" climate.

3) Gigantism in fossil plants and animals

Explanation for Pre-Flood versus Post-Flood life spans of humans.

According to the Bible, humans lived for more than 900 years (at least Seth's line did from those genealogies in Genesis 5). Then, after the Flood, lifespans dropped significantly.

From the study of modern family lines that include centenarians (100-year-olds) and the process of cell division, some believe that lifespans were affected by both environmental factors and genetic ones. Noah still lived another 350 years in the post-Flood atmosphere. The decline reached a plateau by the time of David. Apart from disease and accidents, there is no biological reason why people could not live much longer than they do at present, if they had the appropriate genetic makeup. Something happened in the world that changed things.

Explanation for the Appearance of a worldwide tropical "greenhouse" climate

Geologists admit there's evidence of a worldwide tropical climate throughout the fossil record. Statements from geology books about the past geological time periods include these descriptions: "The earth had a tropical or sub-tropical climate over much of its land surface, & abundance of lush vegetation. The land was low, & there were no high mountains forming physical or climatic barriers." Scientists have found fossils of tropical plants and animals from the North Pole to the South Pole and on top of the highest mountains.

Another indication of warmer climate is coal—compressed plant material that becomes carbonized as all its cells are replaced by carbon. Huge coal seams everywhere represent massive vegetation existing all over the land being buried and decaying at the same time. Nothing like that is occurring today.

Creation scientists propose a pre-Flood world that included these elements:

Tall mountain ranges and ice caps at the poles were absent. Based on Genesis 1:9, we can propose all the land was together in one landmass with fewer and lower mountains. This matches the geological evidence. The combined continents centered around the equator, extending not nearly as far north as today. This would result in a more tropical climate over most of the land as opposed to today. There were likely no deserts yet.

Shallow continental seas trapped heat and provided moderate temperatures to the land. Everything today grows bigger in tropical climates. Yet, a likely explanation for tropical plants and creatures being buried at the poles is that during the Flood, the plants and animals, even live animals, floated to high latitudes from their tropical origins on gigantic floating mats of vegetation. Similar mats are seen during local floods today and after tsunamis scour shorelines.

Larger concentrations of atmospheric carbon dioxide led to a greenhouse effect, increased forestation, and the resulting coal deposits. We do find tremendous amounts of fossilized vegetation everywhere, including current deserts. In spite of adverse publicity about carbon dioxide (CO_2) today, the opposite side is that forests would soak it all up!

Explanation for gigantism in fossil plants and animals.

According to the fossil record, nearly all modern animals and plants were once represented by larger fossil ancestors. Dinosaurs weighing 40 tons (although the average size was that of a sheep). Other examples are a 6-foot tall armadillo (Florida), 12-foot tall bison, 6-foot long beaver, 20-foot tall camel (India), 9-foot tall donkey (Texas), 11-foot long turtle (Texas), and a 17-foot tall rhinoceros. Pigs grew to be the size of cattle. Wombats were the size of a small car. Giant birds nearly 10 feet tall, weighing over 500 pounds, could swallow a dog in a single gulp. Cattails grew to be 60-100 foot tall; club mosses were 120 feet tall; dragonflies had a wingspan of 25 inches; and insects similar to scorpions and centipedes were 3-5 feet long. Yikes!

Giants were in the land! Then all of a sudden, the large animals and plants disappeared. We don't have them today—except for the whale. Why is that? Secular geologists say that the giants have "devolved" to smaller representatives. That doesn't make sense to me. Few of these extinct animals are different species than exist today; most are the same. Scientists solve this problem by the use of the term "like." If a 12-foot St. Bernard were to walk across their lawn, they would describe it as a "dog-like creature" rather than a dog.

Something was different about the pre-Flood world! But what? God created some kind of environment for the earth that produced a worldwide tropical climate and protection from harmful radiation. Was it more water vapor in the air? Was it a stronger magnetic field? Whatever it was, all this resulted in longer life spans for humans and animals.

How did the animals and plants grow so big? We know that many living reptiles and even some mammals (elephants) can grow slowly throughout their lifetimes. A cottonmouth snake can grow to the size of a tire. That's an "old" snake. Comparing bones of juvenile dinosaurs, scientists believe that dinosaurs experienced rapid growth in their youth and then continued slow growth the rest of their lives. If dinosaurs lived hundreds of years, as people did before the Flood, they had plenty of time to develop into the enormous creatures we find as fossils. Maybe some fossils of giants are the ones originally created by God. Also, the climate as described above seems to favor optimal growth of animals and plants. Remember that everything had been created perfectly. There hadn't been much time for the cumulative effects of damaging mutations to show up in genetics. And there is indication from post-Flood evidence that the largest creatures were killed by humans. There were still **giants in the land** after the Flood.

Conclusion

It is reasonable to consider the pre-Flood earth having some kind of greenhouse effect, greater shielding from radiation by both water vapor and the magnetic field, and habitable land space actually being closer to the equator with fewer mountains to block air currents. These plus genetic factors contributing to long lives could easily contribute to not only humans living for a long time, but also gigantism in animals and plants.

What we read in God's Word agrees with what we see in God's World.

For more information

Regarding what we covered in this lesson, go to the Institute for Creation Research at **icr.org** or Answers in Genesis at **answersingenesis.org**. SEARCH a word or phrase from today's passage such as Cain and Abel, pre-Flood civilization, Nod, Nephilim, Noah, pre-Flood, giant animals, etc. Select an article to read or video to watch.

To practice discerning facts from assumption, go to the "Discernment Practice: Interpreting What You Read and Hear" page in the back of this book. Learn how to follow the process given.

Lesson 6: The Flood and the Fossils

The Bible is the infallible Word of God, but study notes in your Bible and commentaries are the interpretations (sometimes best guesses) of educated, though fallible, humans. Like Creation, these may give confusing or inaccurate information regarding the Flood. The best approach is to first study the biblical text, seeing what the Bible actually says both in the passage and elsewhere in Scripture about that event. As we have done previously in this study, ask the question, "If this is true, what would I expect to find?" Then, observe what is found in the world.

DAY ONE STUDY

Ask the Lord Jesus to teach you through His Word. Tell Him you are listening.

Read Genesis 6:8-13.

> **From the Hebrew:** Genesis 6:8 is the first reference to *grace* in the Bible. The Hebrew word translated "favor/grace" (*chen*) is from a root word meaning "to bend or stoop," and thus came to mean condescending or unmerited favor of a superior person to an inferior one.

1. From Genesis 6:8-9, describe Noah's character.

> **Scriptural Insight:** To walk with God refers to being in God's presence, acknowledging Him, and doing life with Him. That is a life based on trusting God and having faith in Him— a faith walk. The book of Genesis describes Enoch (Genesis 5:24), Noah, and Abraham (Genesis 17:1; Hebrews 11:8-10)) as walking with God. They each had a faith walk.

2. From the following verses, what quality of Noah caused him to find favor with God?

    *   Genesis 6:22; 7:5—

    *   Hebrews 11:7 (A similar description of Noah is found in other ancient documents.)—

3. ***Connecting Faith to Sight:*** Noah was the only one in his generation who was found righteous. Isn't it wonderful to know that because of Christ and His Church, a Christian is surrounded by brothers and sisters who are like-minded, also found righteous because of what Christ has done for them? Think of a time when you may have felt like you were the only one who was trying to be righteous and walk with God in your family, workplace, or community. How do you deal with such isolation? What scriptures have given you direction and comfort for such times?

Read Genesis 6:14-22.

4. Describe the Ark that Noah is to build, paying attention to the *minute details* of both exterior and interior layout.

   - Verse 14—

   - Verse 15—

   - Verse 16—

   - Verse 21—

From the Hebrew: The ancient *cubit* ranged from 17.5 to 24 inches, averaging 18 inches. *Ark* comes from an ancient word for "box," meaning to float upon water. It's the same word used in Exodus 2:3 to describe the basket that Moses' mother made for him to float in the Nile River. *Gopher wood* is unknown but speculated to be a hard, dense wood similar to oak or a laminated wood product. *Pitch* means "to cover (verb)" or "a covering (noun)." It's the same word used for "atonement" in Leviticus 17:11. Pitch was probably some resinous substance. It would be the perfect covering for the Ark, keeping out the waters of judgment, just as the blood of the Lamb provides atonement for the soul. Ancient Minoan boat builders likewise coated their boats with a hard, water-resistant resin.

5. Consider the skills and tools needed to carry out God's instructions. Compare the information you learned today to your previous concept of Noah's Ark.

Scientific Insight: Based on the ancient cubit, the dimensions of the Ark would be about 450 feet long, 75 feet wide, and 45 feet high. Hydraulic engineers have determined that the Ark would be exceedingly stable, almost impossible to capsize. When tilted up to just short of 90°, it would immediately right itself again, tending to align itself parallel with the direction of wave advance so as subject to minimum pitching. Its carrying capacity would be at least 1.4 million cubic feet, about the volume of 522 standard railroad livestock cars holding 240 sheep each. So the Ark could carry at least 125,000 sheep-size animals. God's design included three decks, about 17.5 feet high each, with a combined floor area equivalent to ¾ the size of a modern ocean liner. The decks were divided into various rooms (literally "nests"). God called for a window, literally "opening for daylight," that gave light as needed for all creatures inside. There was plenty of room for living quarters for humans, 75,000 air-breathing creatures, and food for everyone for a year or more. Too big for a local flood, no boat of this size was built again until 1884! "The Ark Encounter" exhibit in northern Kentucky is built to represent this size and design of the ark God designed.

Read Genesis 7:1-16.

6. Compare Genesis 1:21-26 with 7:13-16, 21. God groups the creatures He has made on Days 5 and 6 (living creatures) into specific categories, not all of which were represented on the Ark. Marine creatures would survive outside the Ark. How were the air-breathing creatures to be gathered and brought to the Ark?

> **Scientific Insight:** Most land animals are small. If you take twice the known land species today (to account for the extinct ones as well), that leads to ~18,000 species. Even considering that our term species is not equal to *kind*, that would yield around 72,000 animals. These would take up about 60% of the carrying capacity of the Ark. Since the Ark held both wild animals and domesticated ones (Genesis 8:1), God may have sent juveniles of the larger animals because they would need to spend a year without reproductive activity yet be healthy enough to repopulate the earth. Perhaps many hibernated, especially during the dark, stormy times. With all those animals, the manure pile may have been at the bottom of the boat, assigned to the insects.

7. There is a lapse of time between Genesis 6:22 and 7:1. God was patient during this time, waiting for Noah to construct the Ark (along with the help of his three sons).

   • Read 2 Peter 2:5 and Hebrews 11:7. What else did Noah do while building the Ark?

   • Was the Ark big enough to carry other people as well? Did any others respond?

> **From the Hebrew:** Noah is described as a "preacher of righteousness." The Greek word *kerux* refers to a herald, or "one who announces." His labor on the Ark would be a visible message to the people. Add to this any words he announced about what was coming and how they could be saved from it by repenting of their wickedness.

8. Focus on Genesis 7:1-10, 13-16. Notice all the words and phrases repeated. What could God be trying to emphasize to us as we read this part of Scripture?

9. After the animals plus Noah and his family were settled into the Ark, who closed the door, and why is this important?

Scriptural Insight: There was one door. All must enter and leave by the same door. Eight people willingly did so; the rest of humanity did not! Noah preached while he worked on the boat (1 Peter 3:20; 2 Peter 2:5). God, not Noah, closed the door in Gen. 7:16.

10. ***Connecting Faith to Sight:*** Have you thought about the wonderful capability God instilled in humans to be able to build things? Think about how God has gifted you to be creative, productive in your own way. In prayer now, give praise to Him. Dedicate your skills to Him and to His work. Noah chose to be faithful even when he could not visualize the reason for his labor.

DAY TWO STUDY

Ask the Lord Jesus to teach you through His Word. Tell Him you are listening.

Read Genesis 7:6-24.

> ***From the Hebrew:*** The word for "Flood" (Hebrew *mabbul*) is only applied to the Noahic Flood; other floods are denoted by a different word in the original. *Mabbul* is related to an Assyrian word meaning "destruction"; the phrase "a flood of waters" (Genesis 6:17) could properly be translated by "a catastrophe of waters." Similarly, when the Genesis Flood is referred to in the New Testament, the Greek *kataklusmos* ("cataclysm") is uniquely employed. (Dr. Henry Morris, *The Beginning of the World,* p. 107)

11. According to verses 11-12, the Flood was brought on by more than just 40 days of rain. Looking at the 2 physical mechanisms that caused the Flood, what physical effects would each have upon the earth?

    * Springs or fountains of the great deep—

    * Floodgates or windows of the heavens—

> **Scriptural and Scientific Insight:** "Springs" refers to the places where water issues or bursts out of the earth. Other scriptures refer to them existing since creation. "The great deep" means the deepest water—the ocean and subterranean reservoirs. This would include great amounts of water, entrapped below the crust now being released. "Burst forth" means "to cleave, split, break open, break through, to fault" as in the parting of the Red Sea. We see the same thing in Zechariah 14:4, *"the Mount of Olives will be split in two from east to west, forming a great valley."* The phrase is also used in Numbers 16:31 for the ground "splitting asunder." Two times God is said to "split open" rocks or the ground to provide water for His people (Psalm 74:15; Isaiah 48:21). The earth was split, fractured. You would expect great volcanic explosions and eruptions, accompanying earthquakes, and

tsunamis (sometimes called tidal waves). The phrase "windows/floodgates of heaven" conveys the image of a great amount of water previously held back but now being released. *Geshem* in the Hebrew refers to the most violent rain. In Ezekiel 13:11-13, violent rain is described as destroying mortared walls! So the water for the Flood came from three sources: (1) Subterranean fountains of water bursting forth, (2) Violent rain from above, and (3) Waters from the existing ocean spilling over the land.

12. What information is given to us about the waters of the Flood even beyond the first 40 days (vv. 17-24)?

13. Considering the description given and the results of the waters increasing over the earth, what is the author plainly trying to convey to us?

14. Skeptics claim that the Flood of Noah must have been only confined to a local area or regional at most (just involving the Mesopotamian river valleys in modern day Iraq). How would you prove that the Flood was truly global from this passage? For example, think of the implications of having floodwaters covering even a short mountain by ~22.5 feet (which may have been the draft of the Ark).

Scientific Insight: Anyone holding to the view that this was a local or regional flood needs to consider the implications of floodwaters covering even a short mountain by 22.5 feet (which may have been the draft of the Ark). Water seeks its own level. It doesn't flow uphill. To cover a hill 800 feet above sea level near Dallas, Texas, water would also be covering most of Texas and the Gulf Coast, the East Coast, and the West Coast except for the mountain ranges. And that's just a small hill. The Hebrew word *kasah* translated "covered" in Genesis 7:20 means "overwhelmed to the point of washing away." Any hill completely covered by raging floodwaters would certainly be overwhelmed like that. The waters rose and covered the whole earth for 150 days. That's 5 months, way too long for a local flood! Building such a large ark would have been a ridiculous waste of time and resources when God could have sent Noah and his family to a neighboring region. Also, the animals could have just migrated. This was no local Flood. You would be better off saying it never happened than to try to make it local. This had to be a global phenomenon!

Read Genesis 8:1-5.

So far, the Flood narrative has been an account of judgment upon a very wicked world. From this point on, it is a story of **redemption**. After 40 days of violent, heavy rain and water rising from the deep followed by 110 days of it prevailing over the earth, God "remembered" Noah (Genesis 8:1). To "remember" in the Bible is to express concern for someone, to act with loving care for them.

15. What three specific things did God do (in vv. 1-2) on behalf of Noah and his family and all the animals so that the water stopped rising and began to recede?

Scientific Insight: The Ark rested upon the mountains of Ararat (modern Armenia). The mountain known as Mount Ararat is volcanic in origin. It is speculated from studying the glass rocks, salt domes, and volcanic structures found there that the mountain may have been only 6,000 feet in height at the time of Noah's landing. Today Mt. Ararat itself (now 17,000 feet in elevation) abounds in what is known as pillow lava, a dense lava rock formed under great depths of water. The mountain also includes certain sedimentary formations containing marine fossils. It was the highest mountain in the region by far. A rather large number of sightings of the Ark have been reported from explorers on this mountain over the past century, as well as during ancient and medieval times. If found, this would surely be the most important archeological discovery of all time.

16. The water had to have a place to go. Part of Psalm 104 may be referring to the work of God at this time. What information is given in Psalm 104:6-9?

Scriptural Insight: After 150 days, God closed the ocean geysers and restrained the rain so that it was no longer torrential. Genesis 8:3 says the water "decreased," meaning the height had been reached by then and started falling down at that time. God caused the Ark to come to rest on the mountains of Ararat. This is God's grace to protect the occupants from what happened next. God put forces into place to turn the raging waters into receding waters. Genesis 8:1 says He sent a wind. Psalm 104 says that God rebuked the water, implying the waters were a chaotic force to be calmed and conquered. Notice that the mountains rose up, the ocean basins sank. The continents were raised up above sea level as new mountain chains were formed. That requires the release of a lot of lava. Water ran downhill forming new valleys and filling large, deep ocean basins now defined by the nearly vertical continental edges. God altered earth's topography to set the boundary. The evidence can be readily seen in: (1) Continents once below sea level having thick deposits of marine layers. (2) Soft sediments being folded, thrust, overturned and uplifted in a short time. (3) Massive volcanic deposits throughout the rock record. (4) Mass erosional features from receding waters. (5) Larger and deeper ocean basins to hold the receding water. The landforms we see today are a result of the Flood, not Creation or millions of years.

17. From Genesis 8:6-12, what actions did Noah take to determine whether or not the ground was dry enough for them to leave the Ark.

Historical Insight: Scholars have worked out the date for Abram's birth to be 2166 B.C. Taking the lifespans in Genesis 11 from Abram back to the Flood, it can be proposed that the Flood began ~2472 B.C. The Flood began on the 17th day of the second month (600 years after Noah's birth). It wasn't over until 370 days later! That was no local flood.

18. In the entire passage of Genesis 6:12-8:14, count the number of times (in your translation) the words "all," "every," or "everything" are used. Total = _____ times. What is the author plainly trying to convey to the reader?

Think About It: The record of the Flood in Genesis gives every indication of being an eyewitness account, written originally by Noah or his sons. Despite the efforts of many commentators to explain it away as a local flood, it is obvious that the writer intended to tell of a worldwide, uniquely destructive cataclysm. In fact, it would be difficult to imagine how the concept of a universal Flood could be better presented than in the words actually recorded in Genesis. (Dr. Henry Morris, *The Beginning of the World*, p. 116)

19. ***Connecting Faith to Sight:*** God lifted huge mountain ranges out of the earth to make the water drain off the land into the newly deepened ocean basins, leaving behind beautiful landscapes. Understanding this, sightseeing can take on a new excitement and appreciation. We can interpret our world from a Biblical perspective so that it becomes a wonder to us reminding of God's power and majesty. Spend some time meditating on the beauty and majesty of mountains. Have you traveled to the mountains? What do you like about mountains? Do mountains inspire you? If so, how? Feel free to be creative in expressing your thoughts about mountains and landscapes in the space below.

DAY THREE STUDY

Ask the Lord Jesus to teach you through His Word. Tell Him you are listening.

Read Genesis 8:15-22.

The covenant (promise) God made with Noah before the Flood apparently became the standard for ancient governments and contracts after the Flood between a king and his subjects in which sovereign protection is offered in exchange for faithful obedience.

20. Read Genesis 6:13-22; 7:1-9; 7:13-16.

    • What protection did God offer Noah?

    • What was Noah's responsibility?

21. During the Flood, God made good on His promise, carrying out His sovereign protection of Noah. After the Flood, how did Noah show his thankfulness to God for this protection (vv. 18-20)?

> **Focus on the Meaning:** This is the first mention of an altar in the Bible (although we don't know how Abel presented his offering in Genesis 4). The first mention of a concept or practice is always significant to understanding its meaning for the rest of Scripture.

22. Focus on Genesis 8:21-22.

    • Why and how was God pleased?

    • What did He declare?

    • How does His promise affect us?

23. ***Connecting Faith to Sight:*** Regarding the Biblical account of a global flood, scoffers abound. Read 2 Peter 3:3-6. The concept of a local flood, lasting over a year, covering all the local high mountains by a depth of 22.5 feet but not leaving any destructive evidence or affecting the rest of the planet is illogical.

- What was the purpose of the Flood?

- Did God accomplish His purpose?

- What does this tell you about evaluating scripture that describes future judgment upon the earth?

- If you have chosen to place your faith in God's Son Jesus Christ, God has chosen to save (deliver) you from any future judgment upon the earth. How does this make you feel?

Read Genesis 9:8-17. (We will cover verses 1-7 in the next lesson.)

24. God establishes a new covenant with Noah that does not depend upon anyone's faithful obedience. This is an unconditional divine promise.

- What does He promise?

- What is the sign that God gives that He is keeping His promise?

    Scriptural Insight: The rainbow as a sign of the new covenant obviously had significant meaning to Noah. It is possible this may have been a new thing for Noah to see. It was also s a symbol associated with God as seen in Ezekiel 1:28; Revelation 4:1-6; and Revelation 10:1.

25. ***Connecting Faith to Sight:*** Skeptics dismiss the idea of a global flood and claim that Noah's Flood was merely local. Read Genesis 9:11 again. We still have local and regional floods, even in Mesopotamia, that are often very destructive and deadly. Does that mean God has broken His promise? Why or why not? What assurance is there for us today that His promised protection is still operating?

Scriptural Insight: Here are 7 clues from scripture that the Genesis Flood was global and not just regional or local. 1) Size and need for an Ark; 2) Purpose of the Flood to wipe out violence on the earth; 3) Depth of the Flood so that all breathing creatures outside the Ark perished; 4) Duration of the Flood for 371 days; 5) Use of universal terms—all, every, everything used more than 30 times; 6) Promise of no more floods; and 7) The rest of Scripture (2 Peter 2:5; 3:5-6; Psalm 29:10; 104:6-9; Matthew 24:38-39; Luke 17:26-27; Hebrews 11:7). *Conclusion:* It had to be universal and it had to have left lots of evidence!!

Respond to the Lord about what He's shown you today.

DAY FOUR STUDY

Creation Answers—Raging Waters and Buried Dead Things

The most universal of legends is that of a worldwide flood. More than 270 cultures have one. This is unexplainable apart from common experience. Add to that some undeniable observations about our planet that tell us something different happened in the past to change its surface drastically and radically different from what's going on today. Is the Flood just legend, or was it a real event? If true, what would we expect to find. The geologist in me gets excited about this evidence!

Rain, gushing subterranean water, and split continents

From your study of the passage in Genesis 6-8, you know that the Genesis Flood was more than just a 40-day rainstorm. It consisted of subterranean fountains of water shooting out from the splitting and distorting of the crust along with violent rain for months. Waters from the existing ocean rose up and spilled over the land, lasting for 150 days at least. That's a long time!

We now see cracks in the earth's crust circling around the globe like the seams on a baseball. It's a fact. There is no explanation for it. The continents appear to have separated. Massive water-lain deposits of mud and debris cover the earth. Studying the continental shelves and slopes plus drowned canyons extending out from land-based rivers indicate sea level was once several thousand feet lower than it is today. Some parts of the sea bottom have definitely dropped and "seamounts" (non-volcanic, flat-topped drowned islands) cover the middle of the ocean. What would happen when that great amount of water and magma was released from below the crust? Most likely, sinking of the land above it. So the evidence matches the prediction. The model may be accurate after all. The Flood may have really happened just as the Bible said it did.

There is no way that such a lengthy, global flood could have occurred without leaving any evidence! If there had been a year-long worldwide flood, what would the evidence be? **Billions of dead things buried in rock layers laid down by water all over the earth.** What do we find? Billions of dead things buried in rock layers laid down by water all over the earth.

Rock layers that could have formed in 1 year rather than in millions

All that soil, sand, and clay plus rock from the fountains breaking up would be scoured from the surface, worked by the water, and deposited elsewhere. Any evidence? You bet!

Thick sediment layers

Sedimentary layers up to 40,000 feet thick cover the surface of the earth. The question is not whether thick sediments exist, but whether they were deposited slowly over millions of years or rapidly. How rapidly can sediments be deposited?

Some geologists say that 1,000 years are needed to accumulate 5 inches of sediment and then deep burial and thousands of years are needed for cementation to take place. Evidence found in the world does not agree, however. Water is a great sorting agent, especially fast-moving water. Particles of rock and organic matter are sorted by size and density so that particles of similar size are deposited together, giving a layered effect. It is now known that sediment layers do form quickly on river floodplains during floods, in shallow marine areas during storms, and in deep water by turbidity currents as well as the result of other catastrophic events (tsunamis and mud flows). The fountains *burst forth*, not sneaked out, requiring lots of earth movements and massive erosion! Water-laid deposits take a lot of churning, fast-moving water, not necessarily a lot of time!

Extensive sediments over wide areas

You would expect to find thick extensive beds of similar materials extending over wide areas with varying thicknesses. You do find this! One example is the Dakota Sandstone–one of the most distinctive rock layers in the Rockies seen throughout New Mexico, Colorado and beyond. No process today is forming similar deposits. And the Morrison Formation, famous for its dinosaurs, covers parts of 11 states. Thick sandstones reach from the Grand Canyon eastward to the Appalachian Mountains. Names change along the way, as does the thickness, but the layer is the same.

Widespread iron-rich materials

Great thicknesses of red to reddish brown sediments cover North America in a continuous sheet that extends west through the Grand Canyon and eastward to the Appalachian Mountains, commonly several thousand feet thick. The color is due to the presence of hematite, an iron mineral, the color of which is so dominant that a very small amount of it will color the rock a faint shade of red. Volcanic rock contains a wealth of iron minerals. When these are decomposed in water, hematite is formed. Iron comes not only from the weathering of volcanic rock, but also from the associated hot waters released by volcanoes, water that is highly charged with iron. Extensive volcanism during a worldwide flood would release tons of mineral-rich water to mix with sediments as they are deposited, resulting in red coloration.

Either the rock record is the evidence of millions of years, or it is largely the evidence of the Genesis Flood. It can't be both. I think the evidence weighs heavily on the side of the Flood.

Fossils in the rocks buried by Noah's Flood

Rapid burial

A fossil is formed when an organism is trapped and buried very quickly under sediment that seals the organism in and holds its body together. As the sediment turns to rock, so do the creature remains. The soft parts may rot away leaving only hard parts like teeth, bones and shells which are replaced by minerals from the sediments. Or molds of the original creature are left. Fossils must be buried quickly, or they don't form at all. Flooding is the most likely way for an animal to be buried since floods come by surprise, trapping and burying animals.

Everywhere you look is evidence of rapid burial of **billions of dead thing**s buried in rock layers laid down by water all over the earth. There are many examples of fossilized creatures in the process of eating or giving birth, perfectly preserved, which can only happen with rapid and complete burial. Thousands of intricately detailed fossil jellyfish (which disintegrate rapidly) are found in rocks. Myriads of fossils extend through multiple layers (called "polystrate" fossils) that span millions of years of geological ages. For example, a complete preserved skull of an ichthyosaur (an extinct marine reptile) was found buried in a vertical, nose-down position through several layers. Unlike most fossils, the head was preserved in three dimensions, and had not been flattened by the weight of sediment above it. It's impossible except if the layers had been deposited rapidly and deeply to bury the ichthyosaur and keep it from decaying.

Fossil graveyards of mixed creatures

Thousands of marine and land creatures are buried together in great "fossil graveyards," tightly packed in a jumbled heap, choked with sediments, buried before they had time to decay. Fish, birds, reptiles, mammals, insects, and plants—creatures that don't live together—have been piled together and preserved. Whole herds of animals were buried at one time. At Fossil Butte National Monument in Colorado, multitudes of perfectly preserved fish, crocodiles, insects, turtles and palm fronds are fossilized with no signs of decay so they must have been buried rapidly. The most massive, continuous, fossil deposit known on the planet is the Karoo deposit of southern Africa. At least 800 billion vertebrate fossils are buried in a 20,000-foot thick deposit of sandstones and shales stretching out for hundreds of miles. Nothing like this is happening today.

Rapid fossilization

Millions of years aren't necessary for a bone to become fossilized. Researchers have found that chicken bones and wood can be petrified in just five to ten years. Perhaps a dinosaur bone would only take a hundred years to completely mineralize. Besides, not all fossil remains have yet turned to stone. Pieces of the original shell and bone as well as soft tissue are commonly present in sedimentary layers.

Unfossilized bones

Then, there are the unfossilized bones. The most sensational, of course are the dinosaur bones. Scientists from the University of Montana found Tyrannosaurus Rex bones that were not totally fossilized. Sections were like fresh bone and contained what seem to be blood cells and hemoglobin. If these bones really were millions of years old, then the blood cells and hemoglobin would have totally disintegrated. Also, there should not be 'fresh' bone if it were really millions of years old. News reports abound of such fossils still containing collagen or blood.

Sequence of the fossils does not prove evolution

The sequence in which fossils are found does not prove evolutionary development. Rather, it is evidence of sorting and differential abilities to escape floodwaters. Consider that 95% of all fossils are marine creatures, particularly shellfish. That makes sense considering ocean waters swept up from the ocean bottom onto the land. The other 5% is plants and land invertebrates. Less than 1% are vertebrates—mostly fish. Most land vertebrate fossils consist of one bone.

Burial sequence could definitely be influenced by where the organisms lived. You'd expect that large numbers of various marine organisms would be buried together. Above them are the free-swimming vertebrates. Then, at the higher elevations—birds, mammals, and finally humans because the ability to escape would affect burial. Larger creatures can climb or swim; plants can float. Both could be buried later in the process. But that's not all …

Floodwater sorts out objects according to similar shape and size. Streamlined objects settle out first; larger creatures with arms and tails settle out last. Based on the nature of churning water and wave action, there is reason to believe that many initial sediment deposits were reworked and deposited elsewhere. So you can't go by the way they are found in the layers anyway.

It is evident that some areas were definitely uplands during the time of deposition. Sedimentary deposits become thinner in the direction of those uplands (often referred to as crystalline shields).

In fact, many of these interpreted upland areas are completely devoid of sedimentary rock or have very thin layers. According to Genesis 7:20, the highest hills were only flooded by a modest amount of water, likely leaving little room for thick sedimentary deposits to accumulate. This provided a place for large mammals and humans to congregate in an attempt to escape the flood waters. Scientists have identified a pre-Flood land mass across the central United States that is labeled "Dinosaur Peninsula." And it appears that the highest areas—the so-called shield areas—were the most likely locations of human habitation.

Uplift of mountains and plateaus to aid drainage

The continents, once below sea level now with thick water-laid sediments full of marine fossils on them, have been uplifted thousands of feet, folded, thrust and overturned on a gigantic scale. Major faults (due to cracking and shifting of rock) dominate the landscape. Most of the present mountain ranges have been uplifted almost simultaneously and during relatively recent times. Marine fossils are found at the top of the highest mountains, including Mt. Everest and the Andes mountains of South America.

No continents are underwater today for the simple reason that continental crust is made of lighter rock than oceanic crust making the continents more buoyant in the mantle like Styrofoam floating on water. And nowhere can you find flatness of a shallow continental sea for hundreds of miles in all directions as the sediment deposits demand and have it stay that way for supposedly millions of years as pile after pile of sediment is laid down. Grand Canyon is currently 7-8,000 feet above sea level. The land supposedly went up and down several times, remaining almost completely flat. The ocean was over the continent each time. That is harder to believe than considering it a onetime event!

The best answer is that release of water and magma during the Flood would cause readjusting of oceans and continents. As new ocean floors cooled, the rock became denser and sunk deeper into the mantle allowing water to flow off the continents. Movement of the water off the continents and into the oceans would have weighed down the ocean floor more and lightened the continents, resulting in the further sinking of the ocean floor, as well as upward movement of the continents. The collision of the tectonic plates would have pushed up mountain ranges also, especially towards the end of the Flood.

Is there any evidence that this is what happened in a shorter time than is currently taught? The answer is, "Yes!"

- In many mountainous areas, strata thousands of feet thick are bent and folded into hairpin shapes. If the rocks were old at the time of folding and uplift, having been deeply buried and cemented for years, they should behave in a brittle fashion, shattering in the process. But, we see evidence of soft-sediment deformation instead in these tightly bent rock layers. You can see this in the Grand Canyon, the Rocky Mountains, and the Appalachian Mountains.

- During the past, lava evidently flowed much more freely than now, spouting from craters and pushing upward from immense cracks. Massive volcanic activity built up huge blocks of land. The Columbia Plateau in the northwestern U.S. was built up by oozing lava flows stacked one on top of another in rapid succession, several thousand feet deep covering 65,000 square miles. India's Deccan Plateau was formed by a lava deposit that is two miles thick! No doubt, lava poured out of the earth for years after the Flood ended.

Rapidly receding waters left behind distinctive land formations.

The rapidly receding floodwaters have left behind leveled-off land with remnant high spots, terraces from receding water levels, and huge drainage gullies. Initially, rapid withdrawal of water would have scoured wide, flat surfaces across the landscapes though sometimes being deflected around resistant areas forming remnant hills, mesas and buttes. Some examples are: 1) the Llano Estacado or "Staked Plain," an almost featureless surface that stretches hundreds of miles across Texas and New Mexico and 2) mesas and buttes throughout west Texas, New Mexico and other places in the west such as Monument Valley.

The Great Plains in the heartland of America became one gigantic alleyway of floodwaters draining off the continent into the Gulf Coast. Wedge-shaped deposits, thicker on the Gulf end (40,000 feet) and nearly horizontal (tilting 1°), extend from the Dallas area toward the Gulf. A surface rock layer near Nacogdoches is buried 14,000 feet below Galveston. It would have taken more water power than the few river systems located in east and southeast Texas—the Sabine, Trinity, Brazos, and Colorado—to lay down up to 40,000 feet of sediments, thousands of square miles in area! Massive sheet erosion might explain the absence of human and large animal fossils in the remaining Flood sediments. The last to be buried—the floating, less streamlined creatures—would be the first to get eroded away and redeposited in the ocean basins.

Other evidences of receding waters are terraces and remnant valleys. Ancient water lines can be clearly seen along the coasts of all continents. California cities are built on them. In less flat areas, the receding waters split into channels, which cut deep valleys in the soft, unconsolidated earth. Practically all river valleys are far deeper and wider than their present river systems require, indicating they once carried a far greater volume of water. This is more in line with a flood than river erosion. So there's evidence of receding waters. A lot of water, not a lot of time.

Deepened oceans dropped sea level

Studying the continental shelves and slopes plus drowned canyons extending out from land-based rivers indicates that sea level was once several thousand feet lower than it is today. Some parts of the sea bottom have definitely dropped. This is particularly noticeable in oceanic "seamounts" (or, *guyots*)—about 70,000 volcanic, flat-topped drowned islands in the middle of the ocean. The cones have been seemingly chopped off at an average depth of 5-6,000 feet below sea level. Evidently, those volcanoes on the ocean floor grew up to the surface of the lowered ocean where their peaks were eroded and flattened by wave action. Later, sea level rose several thousand feet as ocean basins sank deeper, water continually drained off the land, and glaciers melted. There is no other credible explanation for the seamounts being there.

Mature landscapes rapidly developed

The land surface quickly took on the look of a mature landscape, one that had been there a long time. Observations in the world show that mature landscapes don't take millions of years to develop. Erosion of fresh surfaces takes place very rapidly then slows down as equilibrium is reached with the stability of the surface, plant coverage, etc. Just watch a recent excavation and preparation of land for building. If it rains right after sloping land is smoothed out, gullies will form in the soft land with little effort. Over the next few weeks, if it rains again, the gullies will deepen. But, that will stop as grass and weeds take hold on the unconsolidated earth and as the size of the gully becomes adequate to handle water flow on that slope. It will stop changing. Within a few months, the slope will look like it had been there for years. You've seen that before, haven't you?

Surtsey Island and Mount St Helens

We don't have continents currently going up and down below sea level to observe, but volcanic mountains make excellent laboratories. Within 3-5 months of Surtsey Island's appearance in 1963 near Iceland, the waves had worked the island edges into level, sandy beaches. A few years later, wind and rain had weathered the lava into soil for hollows and glens to become lush meadows. Despite the extreme youth of the growing island, one encounters a landscape so varied that it gives the appearance of being thousands of years old. The same thing has been seen at Mount St. Helens since its massive eruption in 1980. If the bulk of the past geologic activity took place under intense conditions in a brief space of time, it's not unreasonable to determine that mature landscapes could have developed within a few years rather than millions of years.

Caves

Ripe conditions were also available for the rapid development of underground caves—soft limestone, earth movements tilting the land, and groundwater draining off the land dissolving the limestone at a rapid rate. In tropical areas, caves and their formations develop much faster than those in more temperate regions because of higher annual rainfall. Large stalactites are also found growing everywhere—in shopping centers, schools, universities, and even in multi-story car parks. None of those are older than 100 years! One stalagmite in Carlsbad had grown so fast it was able to preserve a complete bat before the creature had time to decompose. A newly formed cave in Mexico has crystals as large as telephone poles that scientists say may have formed in 30-100 years because of the extreme temperatures and humidity. It's possible to consider that caves and their formations resulted more from the after-effects of the Genesis Flood than through slow and gradual processes over millions of years.

Canyons

Then, there are the large canyons on the earth's surface. Nearly all canyons are vastly larger than the river systems they contain. This fact and other recent observations cast new light on canyon formation. A new canyon formed last century near Walla Walla, Washington in less than 6 days due to dammed up water being suddenly released. In less than a week, what once was an insignificant ditch became a gully, then a gulch then a miniature Grand Canyon—1500 feet long, 120 feet deep—carved through silt, sand and solid rock! At Mount St. Helens in one day's time, layers of ancient solid rock, 500-ft thick, were eroded by fast-moving mud flows and water to produce several mature looking canyons 100 feet deep only 4 years after the initial eruption. Later, creeks formed at the bottom of the canyons. The canyons caused the creeks, not vice versa!

The Grand Canyon did get eroded, but probably not by the Colorado River. One current explanation is called the "Breached Dam" theory. There's evidence that large lakes once existed upstream. It is speculated the canyon formed as the natural dams were breached, releasing huge amounts of rapidly flowing water. Later the river developed at the bottom of the canyon. Some geologists are recognizing that the Channeled Scablands of Washington were formed in a similar way as waters breached an ice dam holding back ancient Lake Missoula in Montana. Also, at Yellowstone National Park, a canyon called "The Grand Canyon of Yellowstone" is now said to have been gouged out in one week by the water that broke through an ice dam.

So it is possible to interpret Grand Canyon rocks as having been laid down by the Flood, the area uplifted late in the Flood year, with trapped Floodwaters carving out the canyon itself while draining off the uplifted continents. Interestingly enough, Havasupai Indian legend says that there was a flood over the earth, then the canyon formed by excessive drainage from the flood.

> "One can physically walk along the sequence of rock units that go back before the Flood, then right through the Flood event up until post-Flood times. While the sequence is not complete, nowhere else is such a complete sequence exposed, nor is there so much evidence for the Flood and its catastrophic nature." (Dr. Andrew Snelling, "Rock-solid for creation," *Creation ex nihilo,* Vol. 18 No. 3, page 22)

Conclusion

The fossil record is best understood as the result of a marine cataclysm that utterly annihilated the continents and land dwellers. Genesis 7:23 "Every living thing on the face of the earth was wiped out." You would expect to find billions of dead things are buried in rock layers laid down by water all over the earth. The evidence matches the prediction.

The earth's features appear to have been fashioned largely by rapid, catastrophic processes that affected the earth on a global scale. Could it be the effects of the Genesis Flood? Yes. The evidence matches the prediction. A lot of water, not a lot of time. It was a new world for Noah and his family as well as for the animals.

What we read in God's Word agrees with what we see in God's World.

For more information

Regarding what we covered in this lesson, go to the Institute for Creation Research at **icr.org** or Answers in Genesis at **answersingenesis.org**. SEARCH a word or phrase from today's passage such as Flood water, mountain-building, springs of the deep, the Ark, canyons, caves, fossil graveyards, sedimentary layers, etc. Select an article to read or video to watch.

To practice discerning facts from assumption, go to the "Discernment Practice: Interpreting What You Read and Hear" page in the back of this book. Learn how to follow the process given.

Lesson 7: The New World & the Great Dispersion

DAY ONE STUDY

Ask the Lord Jesus to teach you through His Word. Tell Him you are listening.

Civilization restart

Noah and the other occupants of the Ark bravely stepped out into a very different world from the one they had previously known. They wouldn't have recognized any familiar landmarks. New rivers and lakes needed to be explored and named. The mountains were higher and more rugged looking. The climate was different from what Noah and the animals had known before, with the world only gradually approaching its present semi-arid state. Seeds and cuttings buried near the surface would be sprouting, but the lush forests were gone. There must have been a sense of aloneness on this vast land.

Read Genesis 9:1-7.

1. Compare 9:1-4 with Genesis 1:28-30.

   - What remains the same?

   - What has changed?

   Scientific Insight: How did the animals spread throughout the New World? Well, if God brought the animals to Noah from all over the earth, He certainly could make sure they spread out again through various ways: 1) Animals can rapidly increase in number, migrate, and separate from one another, allowing variation to develop as gene pools are reduced. After the Mount St. Helens eruption in 1980, the decimated elk herd had an increase of multiple births per cow allowing the herd to grow rapidly. 2) As ice formed on the continents, water removed from the oceans lowered sea level to expose land bridges connecting large landmasses and giving animals a means to get to different continents. 3) Animals today catch rides on vegetation mats floating on the ocean surface. When Krakatoa erupted in 1883, the island remnant remained lifeless for some years, but was eventually recolonized by a surprising variety of insects, birds, lizards, snakes and even a few mammals crossing the ocean to get there. So animals migrated to various parts of the earth and established unique populations in single locations. Many became extinct because they weren't suited to a particular climate. Kangaroos didn't make it in Africa, but the group in Australia survived. Hummingbirds didn't make it in Europe, but the ones in North America survived. You wouldn't expect to find fossils of those creatures in other parts of the world unless those fossils had been buried by the Flood. They left the Ark to brave a new world.

2. Focus on Genesis 9:5-6.

   - What does God value?

- What responsibility does He delegate to humans?

- When you compare Genesis 6:11-12 with 9:5-6, what do you think God is wanting to restore to His creation?

> **Think About It**: You can read an excellent article, "Capital Punishment: A Christian View and Biblical Perspective" by Kerby Anderson, at probe.org/capital-punishment.

Read Genesis 9:18-29.

3. Noah was just as fallible as every other human, including drunkenness (vv. 20-21). The ways Noah's sons reacted to their father's drunken state revealed their character. What do you learn about each from their actions?

   - Ham—

   - Shem and Japheth—

4. ***Deeper Discoveries (optional):*** What is God's opinion regarding mocking/disrespecting one's parents? See Exodus 20:12 and Ephesians 6:1-3.

Parents often have great insight into the strengths and weaknesses of their children. Let's look at Noah's prophecies about each of his sons.

5. What did Noah say about Canaan, the son of Ham?

> **Historical Insight:** Because Ham sinned as a son, Noah's curse was directed to Canaan, Ham's son. Like his father, Canaan would experience the dishonor that a son brings to his father. The word translated "servant" or "slave" in the prophecy can also be translated "steward," someone who gives material service to humans. The Hamites as great "servants" of humanity have opened up the remotest parts of the world to settlement (China, India, Africa, the Americas) and developed cultivation of most of the basic food staples as well as structural forms, building tools, and materials. They also developed paper, ink, block printing, movable type, and other accoutrements of writing and communication. Sadly, the Canaanites became a wicked people who rebelled against God by their idolatry. They are long extinct, so the curse cannot be applied to anyone today.

6. What did Noah say about his oldest son Japheth (v. 27)?

> **Historical Insight:** Japheth's name means "enlargement, extension, prosperity." This carries with it the idea of enticing/persuading so it may refer to mental enlargement and open-mindedness. The Japhethites (especially the Greeks, Romans, and later the Europeans and Americans) have stressed science and philosophy in their development. Interestingly, the term "Gentiles" was applied especially to Japheth's descendants.

7. What did Noah say about Shem (vv. 26-27)?

> **Historical Insight:** Shem's name means "honor, authority, character." The prophecy for Shem included the name for God—*Yhwh*, suggesting a love for God and faith in His promises. The Semites throughout history have been dominated by religious motivations centered in monotheism—the Jews, the Muslims, and the Zoroastrians. Through Shem, God's greatest blessing for humanity, the promised Seed of the woman, would eventually come into the world.

Respond to the Lord about what He's shown you today.

DAY TWO STUDY

Ask the Lord Jesus to teach you through His Word. Tell Him you are listening.

Read Genesis 11:1-9.

A couple of years after the Flood, Noah's grandchildren began to arrive. He soon had sixteen grandkids. Genesis 10 and 11 relay what happened to those grandkids and their families. Genesis 9:19 boldly declares that the whole earth was populated from Noah's sons. Chapter 11 tells us what happened and thus shows that these events actually came first. Chapter 10 tells us how (as described in 11:9). It is the genealogy of who they were and where they settled. If this is true, you would expect that the oldest civilization developed in that area, which is exactly what we find.

8. What information is given in verses 1-3?

Historical Insight: Noah's descendants moved east to Shinar (Genesis 11:1-2) and built there. Shinar is the whole region of the Tigris-Euphrates valley—the closest fertile area to the mountains of Ararat. The most ancient peoples leaving written, historical records were the inhabitants of that valley (the Sumerians) as expected, followed by the Nile Valley of Egypt and other near-Eastern areas. This area is called Mesopotamia (meaning "middle"). According to modern computer studies, the geographical center of the Earth's landmass is indeed situated very close to the Mount Ararat region. Coincidence? I think not!

9. What advantages did a common speech give humanity?

Historical Insight: Within 100 years, Noah's descendants were building cities. The Bible says they made brick for stone and used tar for mortar. Since that area has no stone, the people had to make their own rock by way of burning local dirt into bricks. Written records describe the building of Babylon by the same brick-making process as described in Genesis 11:3. Samples of such human-made bricks have been found. One news article said this could only have been made if there were interactions between metallurgists and potters. This news surprised archaeologists that people with different skills would be working together. A second article described how those same ancient people used a sticky, coal-based "tar" as glue on wood and stone, also surprising archaeologists. But not us!!

10. Within 200 years, the population could have grown to several thousand, considering size of families and longevity. What did the people want to do and why (Genesis 11:4)?

Historical Insight: They planned a tower whose top literally is into heaven—meaning representing heaven, a center of worship. We find that the step-like tower known as a ziggurat originated there, acting as an artificial mountain that became the center of worship in the city. Seated in a miniature temple at the top of the tower was their god. Intelligent though foolish, the builders' sin was one of immense pride that led to open rebellion against God, declaring independence of Him. A common language enabled the rebellion. Ancient humans were not backward and ignorant savages as depicted in books and movies. Their forefathers built cities, herded livestock, played musical instruments, made tools of bronze and iron (Genesis 4:17-22), and built an unsinkable boat! Building a city and tower required marvelous engineering skills. We know the ancient Sumerians had electric batteries and did electroplating of metals. The civilizations that shortly followed—Babylon, Egypt, Greece, China, India and in the Americas (Inca, Maya)—were amazingly advanced. This is seen in their elaborate architecture, calendars, writing, organized cities with intricate water supply systems, and amazing understanding of astronomy and seasonal changes. Theirs were awe-inspiring accomplishments without our modern-day mechanics.

11. What did God think of their behavior and decide to do about it (vv. 5-8).

12. **Connecting Faith to Sight:** The Babel builders sought to make a name for themselves by erecting their tower, a foolish thing in the eyes of God. We just studied how God rearranged the plans of the people to bring them in line with His plan. Has God ever done this in your life, and how did you respond?

Historical Perspective

God separated the people by confusing their languages, forcing them to spread out and repopulate the earth. Genesis 10 describes how they spread out by their clans and languages (Genesis 10:5, 20, 31). Archaeology confirms that many civilizations appeared about the same time, only a few thousand years ago. Not surprisingly, the building of the Tower of Babel was carried with them as well. We find pyramidal-shaped buildings in most ancient civilizations around the world, particularly in those who descended from Ham. No human people group today (or in the past) speaks a "primitive" language, even those in isolated cultures. Each language has nouns, verbs, etc. The observable evidence tells us that by 2000 BC all the people had broken up into these different language groups. If this is true, we would expect to find: 1) evidence through archaeology, written records, and language studies that all people have migrated from that area and 2) similarities of cultures in civilizations spread thousands of miles apart, including histories that contained traces of Creation, the Fall, and the Flood.

Two things make it hard to find evidence. Since the 1960's, it is politically incorrect to suggest that a culture did not develop itself independent of others. General archeology is evolution-based, assuming slow and gradual cultural development. The truth is that the entire world could have been populated in just centuries. Considering only 6 children per family, within 250 years you could have a ~100,000 people on earth. Ice Age land bridges connected England to mainland Europe, Asia to North America (the Bering Strait), and the Malaysian Straits to New Guinea and perhaps Australia. The people could have traveled by foot, boat, and by wagon. Migration campsites are found along the way from Asia to the Americas, some indicating they had boats and could do offshore fishing. Since the animals had preceded them by some time, hunters could kill bison, mammoths, mastodons, camels and giant sloths for meat as they traveled. There were wagon wheels 5,000 years ago. The early human presence on South Pacific islands means they were able to build boats and sail. Not surprising, much evidence from antiquity indicates that they did!

Remember that all *recorded* history has taken place in the past 5,000 years or so. Before then is speculation. People, as the only reliable recorder of history, were already writing on clay tablets and scrolls (books), inscribing with an iron stylus on lead, and engraving in rock slabs. In fact, Job 19:23-24 confirms this shortly after the Tower of Babel time. Truly verified archaeological dating (based upon written records, not radiocarbon dates) do not predate the time of the Biblical Flood, between 3000-2350 B.C. *Just as expected!*

Respond to the Lord about what He's shown you today.

DAY THREE STUDY

Genesis 10 is called the Table of Nations. It is an astonishingly accurate document and unique. Most of the nations disappeared from the historical scene centuries before Christ was born, making it a very ancient document indeed. Many of these early nations kept an accurate record of their beginnings and wrote down the names of their founding patriarchs, providing us with a surprising link between the ancient post-Flood era (Genesis 9-11) and the rest of recorded history.

The first generations after the Flood lived to be very old. For example, Noah lived 350 years and died only 3 years before Abraham was born. People often called themselves, their land, and often their major city or river by the name of the man who was their common ancestor. Sometimes the various nations fell into ancestor worship making it natural for them to name their god after the man who was their ancestor or to claim their long-living ancestor as their god. By using these clues, we can trace most of the nations that formed from Noah's sons.

Use the map below to see where each of Noah's descendants settled.

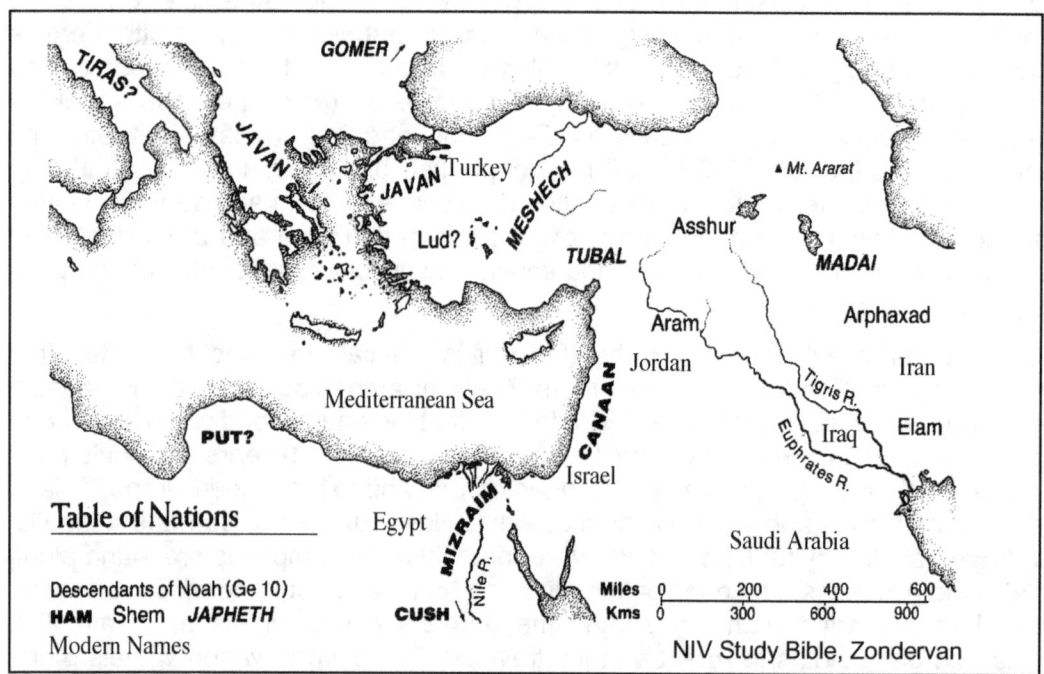

Ask the Lord Jesus to teach you through His Word. Tell Him you are listening.

Japheth and His Kin

Read Genesis 9:18-19; 10:1-5; Acts 17:26.

13. On the map, locate the names of JAPHETH'S DESCENDANTS mentioned in this passage. Color the area around each one blue.

Historical Insight: Six of Japheth's sons headed north to Europe (eventually including Great Britain, Ireland, and the Scandinavian countries) and western Asia; 1 headed east to Persia and India. Noting pertinent references in Scripture, early secular historical writings, and on excavated archaeological monuments, we can trace most of these nations and recognize them as ancestors of the Indo-European peoples, a classification based upon common language. The Indo-European family of languages was/is spoken in nearly all of Europe and much of India, Pakistan and parts of western Asia. Here are some fun facts about Japheth's legacy:

1) The recorded ancestries (very ancient, predating Christianization) of the early Britons, Irish Celts, Anglo-Saxons, Danes and Norwegians all include their descent from Noah through Japheth. The Saxons considered Japheth, whom they called Sceaf, the son of Noah, to be the founder of the European nations and trace their own ancestry through his oldest son, Gomer.

2) The Armenians traditionally claim to be descended from 2 of Gomer's sons—Togarmah and Ashkenaz. Some of Ashkenaz's clan later migrated to Germany, giving their name to that nation since Ashkenaz is the Hebrew word for Germany, and German Jews are called the Ashkenazi. This group of Germanic peoples spread into Denmark (the Anglo-Saxons) and the northern islands.

3) When first encountered by Christian missionaries, the Maiutso People of China already claimed descent from Japheth through Gomer. Their histories (recorded in the form of ancient couplets recited at all public occasions through the ages) included surprisingly accurate accounts of Creation and the Flood (with names of Noah, his wife and 3 sons) plus a graphic account of Babel and the confusion of tongues that resulted in nations spreading out and encircling the globe.

4) In Greek literature, Japheth's name is preserved as *Iapetos*, the son of heaven and earth, the father of many nations, and the legendary father of the Greeks. The ancient Romans perpetuated his name as *ju-Pater* (meaning "Father Jove"), which was standardized to *Jupiter*. Here's an incidence where the ancestor became the chief god.

5) The name Japheth is found in the ancient literature of western India and Pakistan as *Pra-Japati* (Father Japheth), the reputed ancestor of the Aryans of India (Aryan is a form of the word "Iran"), deemed to be the sun and lord of creation for those descended from him.

14. ***Deeper Discoveries (optional):*** Read the following verses to see how Japheth's descendants are represented in them. Isaiah 66:19; Ezekiel 27:13; 32:26; 38:2-6; 39:6; and Revelation 20:8.

Ham and His Kin

Read Genesis 10:6-20.

15. Using the map again, locate the names of HAM'S DESCENDANTS mentioned in this passage. Color the area around each one yellow.

Historical Insight: Of Ham's four sons, some stayed in Babel and the area of Assyria. Most went south into Arabia or west and southwest to Canaan, Egypt, and the rest of Africa. The Bible often refers to Africa as the land of Ham. Yet, Canaan's descendants also headed towards China, eventually populating the Americas through crossing the land bridge that connected Asia with Alaska during the Ice Age. From Alaska south were two corridors free of ice and open to travel. *"Afterward the families of the Canaanites were spread abroad* (Genesis 10:18)." Here are some fun facts about Ham's legacy:

1) Archeologists have noted a number of similarities between the Hittites (possibly being from Canaan's son Heth, Genesis 10:15) and the people of Mongolia (the art of smelting and casting iron plus the breeding and training of horses). Native Americans have their ancestry in the Mongolian peoples. The Biblical Hittites were known to the Sumerians as Khittae and Assyrians as Khatti which may be the origin of Cathay (the English translation of the tribe visited by Marco Polo in China).

2) Biblical and secular references also mention a people in the Far East named "Sinim, Sinae" suggesting some of the descendants of Canaan's son Sin migrated eastward. The Assyrians knew the Biblical Sinites as the Siannu. It is significant that China and the Chinese people have always been identified by the prefix "Sino-" (e.g., Sino-Japanese War), and the name "Sin" is frequently encountered in Chinese names. Recent research of the genetic makeup of Chinese people shows correlation with the people in Africa.

3) Research reported in the late 1990s showed all Native Americans have 4 special genetic markers that appear to trace back to modern Siberians and Mongolians.

4) About 3,000 years ago, the first civilization developed in the Western Hemisphere—the Olmec culture in Mexico and Central America— around the same time that the Shang Dynasty of China fell to enemies, forcing thousands of refugees to flee across the "Eastern Sea." Researchers have recently recognized at least 100 ancient Chinese language symbols embedded in the motifs that decorate Olmec art! The Shang and Olmec societies shared much in common including transportation skills, worship of earth and ancestors on similarly built mounds, human sacrifice, veneration of the same animals, and art carvings.

16. In Genesis 10:8-12, what information is given about Nimrod as a man and the nation he founded?

Historical Insight: The Bible tells of the attitude of men before the Flood in Job 22:15-17, *"Will you keep to the old path that evil men have trod? They were carried off before their time, their foundations washed away by a flood. They said to God, 'Leave us alone! What can the Almighty do to us?'"* That attitude prevailed in some of Noah's descendants in Babylon as well (Genesis 11:3-4). The long war against God gained steam from Babylon outward. Ham's grandson Nimrod is often credited with instigating the Great Rebellion at Babel and founding the worst features of paganism, including the practice of magical arts, astrology, and human sacrifice. Worshipped from earliest times (by Sumerians, Babylonians, and Assyrians), Nimrod was worshipped by the Romans as Bacchus (son of Cush). Various sites in the ancient world were named after him. Shinar was known as the Land of Nimrod. Ancestor worship also began in this area and, combined with other aspects of paganism, was carried by tribes forming new nations—Assyria, China, the Saxons—resulting in the spread of polytheism and idolatry. Degrading from there, the forces of nature were given names (pantheism) so that some people considered everything to be a god—Mother Nature, Father Time, natural selection (evolution itself). As God said in Romans 1:25 "they exchanged the truth for a lie."

17. ***Deeper Discoveries (optional):*** Read the following verses to see how Ham's descendants are represented in them. Isaiah 18:1-2; Jeremiah 46:9; and Ezekiel 30:1-9; 38:1-6.

Shem and His Kin

Read Genesis 10:21-32.

18. Using the map once more, locate the names of SHEM'S DESCENDANTS mentioned in this passage. Color the areas around each one red.

19. Peleg's name means "division." In Genesis 10:25, what information is given about the time period in which he was born?

Historical Insight: Shem's 5 sons headed east, north and south from Babel. The Elamites settled east of Babel, later merging with the Medes to form the Persian Empire. Asshur's name is preserved in his descendants, the Assyrians, whose culture also included the Hamites (Nimrod and company). Unfortunately, Asshur was one of the earliest men to be deified and worshipped by his descendants. Here are some fun facts about Shem's legacy:

1) Arphaxad is the father of the Chaldeans, and his descendants were adept astrologers, magicians, and mathematicians. He is better known by his son, Eber, whose name is possibly preserved in his descendants—the Hebrews.

2) Arphaxad's grandson, Peleg, may have been born shortly after the time of the dispersion. According to *Genesis 10:25, "in his days the earth was divided"* which combined with Genesis 10:5, 32 clearly refer to a linguistic and geographical division.

3) Aram's name is the Hebrew word for Syria. His sons became the Arameans, equivalent to Syrians. Their language was adopted as the trade language for the ancient world, commonly spoken even among Jews in Jesus' day. Some Arameans migrated to Arabia where Job lived.

20. ***Deeper Discoveries (optional):*** Read the following verses to see how Shem's descendants are represented in them. Jeremiah 49:34-36; Numbers 24:20-24; and Ezekiel 27:23-24.

Shem's Descendants—One Branch

Read Genesis 11:10-26.

21. What do you notice about the life spans of the men listed in this genealogy compared to those listed in Genesis 5?

22. Whose ancestry in particular is listed in these verses (see v. 26)?

23. ***Connecting Faith to Sight:*** You descended from at least one of Noah's sons, perhaps all three. From which nations of people have you descended? What do you appreciate about your heritage? Draw your family tree back through four generations. Which ones were Christians and might have had a positive influence on your life? How?

Scientific Insight: What does the Bible say about the races? A total of 70 "families" forming nations is mentioned in Genesis 10. As they migrated out from Shinar and became individual nations, each group developed its own unique physical characteristics. These three streams of nations are *not* three "races." The Bible never talks about races but about tribes and nations. We are told in the Bible that we are all of one blood, one flesh (1 Corinthians 15:39). There is really only one race—the human race! The modern concept of "race" is based on evolutionary thinking.

Modern genetics shows that when a large, freely interbreeding group is suddenly broken into many smaller groups which from then on breed only among themselves, dominant characteristics unique to each group will arise very rapidly. There are four genes for skin color from which a whole range of skin shades may be produced in the offspring from light to dark. All skin color in humans comes from a pigment called melanin. We all have the same color; some just have more of it than others (intensity). A person with a lot of melanin will be very dark; a little melanin renders her very light. There are dusky and black people found among all three groups of nations. All the information for the variations of skin shades goes back to the original parents—Adam and Eve.

Respond to the Lord about what He's shown you today.

DAY FOUR STUDY
The Ice Age, Dinosaurs, and a Corrupted Witness to the Truth

This article will attempt to give scientific answers to three questions related to life in the New World after the Flood.

Was there really an Ice Age?

Besides all the changes in the landscape as floodwaters were draining off the land, the climate would have changed as well—at first in the polar regions, later among the tops of the new tall mountains. This could have brought on the "ice age."

An "ice age" is a time of extensive glacial activity that covers a relatively large area with ice. During the Ice Age, which ended a few thousand years ago, 30% of the land surface of the earth was covered by ice. We know this from the deposits left behind by the advancing and retreating ice sheets plus other surface features. Today's polar ice caps and alpine glaciers are the remains of those vast ice sheets. The evidence points to just one ice age, though.

To accumulate and maintain large sheets of ice requires 3 conditions: 1) increased evaporation, 2) increased snowfall, and 3) decreased snowmelt. All three requirements are met by the Flood model, especially through extensive volcanic activity during and after the Flood and the residual warm oceans following the Flood.

Increased evaporation

For increased evaporation, the key is more heat in the ocean. The massive amounts of hot water and magma released from below the ocean heated the ocean water, rendering a much higher level of evaporation than that in the modern cool ocean. Under such conditions, most of the resulting snow would fall in the middle latitudes and polar regions. Studies of ocean floor sediments show that the oceans were at one time 40° warmer than today. This would encourage evaporation and more snowfall available to higher latitudes and elevations. You would expect to find thick snow accumulation forming ice sheets.

Increased snow fall

Increased snow fall is a direct result of warm water evaporating and falling on cold continents. Volcanic dust trapped in the atmosphere following the Flood would have reflected some of the sunlight back to space and caused cooler summers, mainly over large landmasses. Ice cores taken from Greenland and Antarctica show abundant volcanic particles and acids in the sections associated with the Ice Age.

Decreased snowmelt

Slick continents, denuded of vegetation, would radiate heat back to the atmosphere thus decreasing snow melt. Volcanic ash would have reduced incoming solar radiation, enhancing the cooling. The result would be cooler summers, not necessarily colder winters. In 1883, Krakatoa in East Indies sent up 3 cubic miles of volcanic dust into the stratosphere, which was then distributed globally. This caused red sunsets for 10 years and lowered the average temperature of the earth 4°F for about 3 years.

How long did it last?

Creation scientist speculate that the accumulation of ice to its maximum depth (average 2,300 feet) took about 500 years and could have melted within 200 years. Of course, Greenland and Antarctica would have continued to grow because of their high latitude and altitude. The ice sheets seriously affected life in Europe and North America. Interestingly, there's a possible reference to this time in the book of Job.

*From whose womb has come **the ice**? And the frost of heaven, who has given it birth? Water has become **hard like stone**, and the surface of the **deep is imprisoned**. (Job 38:29-30)*

What about the rest of the land? Land close to the warm oceans and on either side of the equator would still have had a pleasantly warm climate like today. Two-thirds of the land was NOT covered by ice. Present day deserts were well-watered. Israel was once a lush land flowing with "milk and honey" producing abundant, large fruit. People once lived in the Sahara Desert along with a variety of tropical and aquatic animals. They have left countless stone tools, pottery, and pictures of those animals carved on the rocks. Explorers have even found fishhooks and harpoons!

If you had all that water tied up in glaciers, the ocean levels would be considerably lower, in fact the sea level was ~600 feet lower at this time. The evidence for this is the drop off from land's edge to the continental slope edges. Because of this lowered sea level, land bridges connected continents and islands, thus assisting the migration of animals and later people after the Flood. Eventually, the seas gradually cooled, so evaporation decreased, therefore the snow supply for the continents would also decrease. As the ash settled out of the atmosphere, more sunlight warmed the earth. So the ice sheets began to melt. Sea level rose to its current level. Rain declined in the middle latitudes so that deserts formed.

Human migration from Babel

All of these climate conditions affected human migration from the Tower of Babel. As the clans grew, individual families separated and migrated farther from one another, their languages gradually changing as well as their lifestyles. Each family group learned how to survive in the land where it had migrated. A pattern seems to have been repeated:

1. A tribe moved to a new area, set up a temporary camp until a more permanent site for their village or city could be established once the area was determined capable of supporting the people—game for hunting, water, and farmland.

2. Temporary homes were made of stones, hides, or sticks; even caves were used. Remains of these original occupation sites naturally suggest to evolutionists a "stone age culture" when actually they reflect only a very temporary situation.

3. Some groups have continued the simple style of living up to the present time; most have progressed to urbanization.

4. As soon as material for ceramics and metals could be found, the "stone age" at the site was succeeded by a "bronze age" or "iron age."

5. Groups farther away from Babel may have lost much cultural 'know–how', including the loss of previously written, as well as spoken, language. Thus, it is not surprising that they would soon develop an alternative style of writing.

Neanderthals were part of Japheth's Indo-European language group who migrated away from the Tower of Babel and found themselves in harsh Ice Age circumstances especially in northern Europe (average temperature estimated to be 14°F lower than now). Some were forced to live in caves where their bones and artifacts are now discovered.

Yes, there really was an Ice Age after the Flood. Next question …

What happened to the dinosaurs?

First of all, the Bible says that God created all the marine and flying creatures on Day 5 (that included plesiosaurs and pterodactyls) plus land animals on Day 6, which included dinosaurs. Human and dinosaur footprints are found together in several little-publicized places.

84

Dinosaurs got on the Ark, likely as juveniles. Those outside the Ark were buried by the Flood. Dinosaurs got off the Ark, multiplied, and spread to all parts of the world through the land bridges. They died out. Lots of animals have died out from this earth, and the memory of them has faded or been corrupted. The name "dinosaur" wasn't even coined until the 1800s several years after the first bones of these creatures were found and studied scientifically.

Were dinosaurs really as vicious as they are depicted in the movies?

Although Genesis 1:29- 30 tell us they were created to be plant eaters before the Fall when creation was cursed by God, enough is known of dinosaurs to strongly suspect that at least some of them ate meat by the time of the Flood. Fish and lizards are found in the stomach contents of some dinosaurs and flying reptiles. Were they vicious? Several types did have sharp teeth, sharp claws, spikes, armor plates perhaps used for offensive or defensive purposes, but many animals today that have sharp teeth use them for strictly peaceful ends—such as the panda and fruit bat. After the Flood, God begins holding the animals accountable for the lifeblood of humans so that flesh-eating dinosaurs may have been punished by God through death. But they were not forbidden to kill other animals for food.

All the nations of antiquity wrote about dinosaur-like animals called *dragons*. The Biblical writers, inspired by God, wrote about creatures described by the Hebrew word *tanniyn* in over 20 passages. They are depicted as dwelling in different types of habitats and being of various sizes. Up until the last century, that word was translated into English as "dragon" but is now usually translated as "jackal" (Isaiah 34:13; Micah 1:8; Malachi 1:3). It is possible that leviathan and behemoth described in Job 41 were dinosaurs of some kind.

How long did they live after the Flood?

Evidently the dinosaurs gradually died out or were hunted to extinction. Their existence has been preserved for us by people all over the world through the many legends of dragons, the descriptions of which closely resemble dinosaurs. Dragon legends are best understood as the faded and/or embellished memories of real human encounters with such beasts. Dinosaur-like animals have been drawn, written about and told about since the beginning of recorded history on every continent. People in India worshipped huge hissing reptiles they kept in caves. In the 10th century, an Irishman wrote of his encounter with a large beast resembling a *Stegasaurus* with "iron" nails on its tail that pointed backwards and a head shaped like a horse. An ancient carving in an Angkor Cambodian temple also looks like a *Stegasaurus*. A Scandinavian legend describes a reptile-like animal that had a body the size of a large cow, 2 long back legs, 2 very short front legs and with very large jaws. Perhaps a T-Rex!

We also know about them through drawings left behind that look very much like known dinosaurs or composite drawings of various types. An ancient Native American drawing found on a Grand Canyon wall looks very much like an *Edmontosaurus*. The ancient Incas sketched creatures that look just like *Triceratops* and other dinosaur creatures. Thousands of dragon stories and pictures can be found in ancient Chinese books and art. Chinese kings used dragons for pulling royal chariots. In the 1500's, a European scientific book listed several animals, which to us are dinosaurs, as still living back then, though relatively small and extremely rare.

We know about them through written accounts. Flying reptiles have also been seen and described in written form by respected writers (Herodotus, Aristotle, Strabo) in ancient Egypt, Ethiopia, India and Arabia. In 1856 France, a flying reptile was blasted out of a tunnel and died. It was studied by a scientist, declared to be a pterosaur with a 10' wingspan. Stories of giant man-eating birds are common among many Indian tribes of the American Southwest. In Utah's Black Dragon Canyon there is a beautiful pictograph of a pterosaur. These were called "thunderbirds."

The most remarkable descriptions of living dinosaurs are those that the Saxon and Celtic people of Europe have passed down to us. The British Isles have historically been the habitat for lots of reptilian monsters. Over 80 places have been recorded where men, women and children have personally encountered them—not just in the past 100 years and not just in Loch Ness (or any other Scottish lake from which have been numerous sightings). The written accounts of many of these creatures include detailed descriptions of their skin and scale colors and refer to multiple witnesses. One church at Llanbadarn-y-Garrag in Wales contains the carving of a local giant lake-dwelling reptile whose features include large paddle-like flippers, a long neck and a small head. One refers to a creature that spat poisonous venom at its victims. Another describes the creature as "being vast in body, with a crested head, teeth like a saw, and a tail extending to an enormous length." It liked to eat sheep, and shepherds. Local archers tried to kill the beast, unsuccessfully due to its impenetrable hide. At Carlisle Cathedral in England, a 15th century bishop's tomb has brass carvings of creatures that look like dinosaurs.

The Saxon poem *Beowulf* describes 3 specific creatures: a flying reptile (the description of which matches a *pteranodon*), a sea dragon (common in harbors, depicted on Viking ships) and a land monster called a *grendel* that was larger than a human, bipedal with small and puny forelimbs, and slayed with its mouth or jaws—swallowing the body of its victims rapidly in large gobbles! A large carved stone in an old Saxon church (in England) shows such a creature biting the necks of other quadrupedal creatures. Woodcut illustrations in medieval European books also show bipedal, scaled reptiles with large mouths. Too many incidents like these are reported down through the centuries and from all sorts of locations for us to say that they are all fairy-tales.

What happened to them?

They died. Lots of animals have died out from this earth. The world after the Flood was much different than before. The environmental conditions with the sparse vegetation and the temperature extremes during the ensuing Ice Age would have caused many animal types to become extinct, a process which continues today. Hunters wiped out mammoths and mastodons; perhaps they wiped out dinosaurs as well. Had there been zoos, maybe some could have been saved. In Wales, the flying reptiles were reckoned to be as bad as foxes for poultry so were routinely killed. Various accounts describe how the marine reptiles hanging around harbors were hunted and killed.

Most creationists believe that dinosaurs have coexisted with humans from the beginning, only becoming extinct in the Middle Ages. Evolutionists have difficulty explaining away dragon legends, cave drawings, and even modern sightings. They completely ignore human footprints in dinosaur strata and even inside dinosaur footprints. They can't explain the evolution of dinosaurs or their demise. All they can do is get you excited about their imaginative recreations of a world that only exists in their minds.

Conclusion: Dinosaurs did not live millions of years ago; they did live beside people; they did go on Noah's Ark; they did live after the Flood; and they are possibly mentioned in the Bible (Job 40-41). There is much evidence that they have lived until quite recently. God created all animal types. Dinosaurs did not evolve, and there is really no mystery about what happened to them.

It is important to defend the book of Genesis to our children and ourselves. After all, if the first book in the Bible can't be trusted in our eyes or in their eyes, why should any other? As one woman once asked, "When my church told me that I had to accept evolution, and that Genesis couldn't be believed as written, I must then ask, 'When does God start telling the truth?'" Our God tells the truth from the very first verse!

Are legends a corrupted witness to the truth?

Popular thought supposes that the nations of the world only became aware of the God of Genesis after they were populated by Jews or evangelized by Christian missionaries, not considering the possibility that pagan humans were indeed aware of God and His attributes, and that this awareness had existed and flourished for centuries without any exposure at all to the scriptures.

As people spread out, they took the story of Creation, the Curse, and the Flood with them. Flood legends (over 300), predating any contact with outsiders (especially missionaries), are found in almost every culture around the world, about 30 of them in writing. Some are remarkably close in their details to the story told in the Bible. Of those legends, 66% say that the Flood was caused by the sinfulness of man, 95% say that the whole world was covered by water, 86% say that there was a favored family which was told to build a boat so that the animals could be saved. *Just as expected!*

The details are different, but the essence is the same. It becomes clear that the original is, in fact, the Bible story.

Here are some examples:

- An ancient Greek document *The Sybil* records this: "When all men were of one language, some of them built a high tower, as if they would thereby ascend up to heaven; but the gods … gave everyone his peculiar language … After this they were dispersed abroad, on account of their languages, and went out by colonies everywhere … took possession of that land which they lighted upon, and unto which God led them … There were some also who passed over the sea in ships, and inhabited the islands."

- The *Irish Chronicles* reckon the dates of historical events by counting the years since the Creation (4000 BC). Two versions of the *Anglo-Saxon Chronicles* say that 5200 years covered the time from Creation until the year AD 6.

- The epic poem *Beowulf* contains many references to Genesis 1-6 (including Cain, giants and a flood) with an abrupt cutoff at that point. Beowulf himself (a historic person) was a descendant of Saxon kings who claimed their ancestry from Noah through Japheth.

- The Toltec Indians of ancient Mexico record a story of the first world that lasted 1,716 years and was destroyed by a great flood that covered even the highest mountains. Their story tells of a few men who escaped the destruction in a "toptlipetlocali," which means a closed chest. Following the great flood, these men began to multiply and built a very high zacuali," or a great tower, to provide a safe place if the world was destroyed again. However, the languages became confused, so different language groups wandered to other parts of the world. The Toltecs claim they settled in Hue Tlalpan (southern Mexico) 520 years after the great flood.

- Likewise, the Mayans of Central America measured time from the beginning of their culture after the Flood ~3113 BC.

- The ancient Chinese preserved the truth about humanity's earliest history in their writings (~2000 BC), referring to a violent catastrophe that happened to the Earth. The entire land was flooded with water reaching the highest mountains, completely covering all the foothills, and leaving the country in desolate condition for years afterwards. This history records that one man, his wife, 3 sons, and 3 daughters escaped a great flood, were the only people left alive on earth, and repopulated the world afterwards. The accounts of Creation, the Fall, and the Flood are revealed in ancient Chinese characters (which combines pictographs to form new ideas) on early artifacts. This shows that these memories were well-established and already exported to other Asian countries before the earliest Christians influenced China ~640 AD.

What we read in God's Word agrees with what we see in God's World.

For more information

Regarding what we covered in this lesson, go to the Institute for Creation Research at **icr.org** or Answers in Genesis at **answersingenesis.org**. SEARCH a word or phrase from today's passage such as Noah and his sons, Nimrod, Tower of Babel, languages, Neanderthals, races, Ice Age, dinosaurs, Flood legends, etc. Select an article to read or video to watch.

To practice discerning facts from assumption, go to the "Discernment Practice: Interpreting What You Read and Hear" page in the back of this book. Learn how to follow the process given.

The Rest of the Story

God made many promises of hope to His people and gave many clues so His people would recognize the One embodying that hope. While some of Shem's descendants headed east and north, many stayed in the Mesopotamian area. Among these were the family of Terah, the father of Abraham and Sarah (the half-sister and wife of Abraham). Through the line of Shem, God's greatest blessing for humanity, the promised Seed of the woman, would eventually come into the world.

God's work of restoration has begun with people first. Every believer who is alive on planet earth or has been alive since Jesus' resurrection experiences the wonderful blessings of both a restored relationship and a restored status with God. The rest of creation (the earth itself including the animal kingdom as well as the entire universe) are still under the curse of decay and death. Romans 8:19-22 describes the whole creation groaning until the time when it, too, will be liberated from bondage.

Some of our Bible Studies cover how God has worked and is still working to provide restoration to us of what was once lost.

- *Everyday Women, Ever-Faithful God*—covers Old Testament women through whom the Messiah descended including Sarah, Rahab, Ruth, and Bathsheba.

- *Profiles of Perseverance*—covers the lives of Joseph through whom the Israelites migrated to Egypt and David, the king of Israel and ancestor of Jesus.

- *Reboot Renew Rejoice*—through 1 and 2 Chronicles, see how God perpetuated the line of King David through whom would come Jesus

- *Identity: A Study of Ezra, Nehemiah, Esther, Daniel, Haggai, Zechariah, and Malachi*—seeing how God preserved Israel after the Babylonian Exile so that it would be ready for the coming of Jesus several hundred years later

- *Heartbreak to Hope*—Jesus' life as recorded in the gospel of Mark

- *Graceful Living*—what Jesus' death on the cross and resurrection accomplished for us to have a restored relationship with God forever

- *Perspective*—studying 1 and 2 Thessalonians shows what God has planned for our future and especially the restoration of creation

You can find all of these Bible Studies and more at melanienewton.com.

Discernment Practice: Interpreting What You Read and Hear

So what are the facts in a news report? What are the assumptions? Separate the two. What did they really find? When what is presented as facts doesn't agree with the model, there may be a problem with the model or the facts (how they were derived).

We believe that God's Word is true. So if the facts to support it aren't available yet, we are confident that they may one day be available. Let's practice examining what is presented as scientific fact and how it fits with the biblical creation model.

So much information that is proclaimed in the media (news, magazines, movies, internet, and books) is biased against the biblical truth presented in the Bible. Often, there is a morsel of fact mixed with a ton of assumptions that are presented as facts, even if they are pure speculation. Every believer needs to be able to discern the truth from speculation.

Follow the steps below to separate the actual facts being presented from biased interpretations or assumptions being stated as fact.

- Look for the facts. Discern facts from interpretation. Mark the facts.

- Recognize the interpretations or assumptions. Question the assumptions used to derive the facts and/or interpretations.

- Consider the same facts with the assumption of Biblical history.

In the following articles, the facts are marked in grey highlighting; interpretations and assumptions are underlined.

#1 "Chilled Crawfish" by Alexandra Witze, Dallas Morning News, 7/13/1998

Ordering crawdads on ice has a new meaning now that geologists have found the oldest fossil crawfish ever—in Antarctica. The discovery of the 280 million-year-old crustacean suggests that the climate around the South Pole was relatively balmy at the time—at least in the streams and lakes where crawdads live …

The scientists found just a fragment of the ancient crawfish in the Transantarctic Mountains in the central part of the continent. The fossil was a part of a claw that had been gnawed on by a predator, perhaps in search for a primitive etouffee. The researchers also found large burrows, dated to about 240 million years ago, that had been dug by crawfish in another part of Antarctica. The burrows indicate that crawfish had evolved their social activity much earlier than scientists had thought. The fact that crawfish could live in the lakes of Antarctica at the time supports other evidence that Antarctica was experiencing a warming trend at the time, the scientists wrote.

- Facts: A portion of a claw identified to be from a crawfish in Antarctica.

- Assumptions: The age of this fossil piece, the social structure of the animals that it represents, and the presence of burrows.

Comment: The crawfish did exist. It was likely buried during the Flood. Structures identified as burrows may or may not be real. They could be escape attempts from creatures being buried by sediments in the Flood. The warmer South Pole fits with the expected pre-Flood climate.

#2 "Arctic Redwood" from GSA Today, January, 2002

'Spectacularly preserved' *Metasequoia* wood has been found at the Fossil Forest site of Axel Heiberg Island (Canadian High Arctic). 'Some of this stuff looks about like driftwood on the beach, but it's 45 million years old,' said one researcher. 'These fossils are chemically preserved at a level you usually would expect to see in something that's only 1,000 years old.'

Comment: The fossilized wood looks young but is given an old age because of evolutionary bias.

#3 "Bees' Nests Baffle Boffins" from Daily Telegraph (London), May 26, 1995

Ancient bees' nests in Arizona's Petrified Forest are baffling evolutionists. How, they ask, could the bees have survived for more than 100 million years before the evolution of the flowers they fertilize and depend on for nectar?

Paleobiologist Dr. Stephen Hasiotis, from the US Geological Survey, discovered that fossilized logs in the Petrified Forest are riddled with holes which he believes are the nests of bees. The logs with the nests are dated at 220 million years old on the evolutionary time-scale, while the flowers that provide nectar for bees allegedly date from only half as long ago. Evolutionists are forced to conclude that either flowers appeared earlier, or the first bees did without flowers for a long time.

*Comment: Bees did not have to wait 100 million years because these vast ages are only **interpretations** of the evidence, not facts.*

#4 "Evolution in Pollution?" from ANSTO, www.ansto.gov.au, 17 September, 2002

Under a headline 'Darwin's Theory Holds True in Northern Territory,' the media release from the *Australian Nuclear Science and Technology Organisation* explained that, 'thanks to 40 years of evolution,' fish in Australia's Finnis River have learned to live with copper pollution from mine wastes.

'Each year a majority of rainbow fish were killed in the first flush of heavy metals downstream at the start of each wet season,' said one researcher. 'However, the few remaining fish passed their ability to survive onto their offspring.' But the downside is that the adaptation to high levels of pollution 'may have occurred at the expense of some other traits that are important for their survival.'

*Comment: While the headline might shout that evolution is true, closer reading shows that it was not evolution but simply natural adaptation. Fish which already had the genetic makeup to survive copper pollution were present in the population, i.e. no new genes evolved. And in common with many similar examples of adaptation to some new harsh external factor, 'at the expense of some other traits' suggests a mutational **loss** of information.*

#5 "Iiwis Reduce Their Bill" from Nature, Volume 375, May 4, 1995

A Hawaiian bird is evolving a shorter bill, evolutionary researchers claim. They say that the Iiwi (pronounced *ee-EE-wee*), a honey-eater with a long, down-curved bill, has changed its eating habits because its main food source, the lobelioid flower, has become rare.

The Iiwi's long bill is ideal for extracting nectar from the base of the deep corollas of lobelioids. But with the large-scale disappearance or extinction of several species of the flowers, the Iiwi now feeds largely on the flowers of the Ohia tree. Other honeyeaters that feed on Ohia flowers have short bills, because these flowers lack corollas. Researchers who compared the size of living Iiwis' bills with the bills of museum specimens collected before 1902 show that the length of the Iiwi's bill has shortened 2-3 percent. Other features have remained stable.

Comment: A variation in beak size is not evidence of bird evolution. Such minor changes merely offer better survival advantage, but are no evidence that birds evolved from non-birds, as evolutionists claim.

#6 "'Superbug' Did Not Evolve" from Nature, 1 August, 2002, p. 469

A new strain of *Staphylococcus aureus* which is resistant to the antibiotic vancomycin has been found in a hospital patient in Michigan, USA. DNA sequence analysis revealed that the new strain did not 'evolve' resistant genes, but acquired them by gene transfer from relatively harmless gut bacteria called *enterococci*, carried by the same patient.

*Comment: That is, the information was already present. In every case known, antibiotic resistance is **never** the result of new genetic information evolving into existence.*

#7 "Evolutionary Back-Tracking" from Science, 10 May, 2002, pp. 1112–1115

Fossil tracks from a sandstone quarry in Canada—made by a 30 cm (1 foot)-long arthropod (a segmented animal with external skeleton, like a millipede)—have astounded experts because they are in Cambrian rock, allegedly 544 million years old. The date at which sea animals came ashore was once claimed to be the Silurian, which allegedly started 440 million years ago. Then it was the Ordovician, which started 490 million years ago, and now this evidence pushes it 'back' another 40 million years.

The date at which many-celled animals first arose has also been pushed back. Current theory says it was 600 million years ago. But now impressions in sandstone rocks of Western Australia that appear to be the trails of an earthworm-like creature have been 'dated' at over twice that age, from 1.2 to 2.0 billion years old. For evolution-believers, this leaves even less time for evolution to have supposedly wrought its work.

Comment: Evolutionary 'dates' are really just relative positions in the strata that were mostly laid during the global Flood. Dates getting 'pushed back' simply means that a fossil has been discovered at a lower level. Thus, we can expect this phenomenon to keep recurring.

#8 "Neanderthal flute?" from The Sydney Morning Herald, February 21, 1996.

The earliest known flute has been discovered in Slovenia in south-east Europe. Archaeologists in the former Yugoslavian republic claim the 12-centimeters (5-inch) flute was made by Neanderthal humans 45,000 years ago. The instrument was made from the leg bone of a bear, and its original four finger-holes are intact. Its lowest note was identified as a B flat or A.

The flute was found in a cave near the town of Nova Gorica, 65 kilometers (40 miles) west of Solvenia's capital, Ljubljana. The revelation that Neanderthal Slovenes learned to play music is said to have 'far-reaching implications for human evolution.'

Comment: Ignoring the 45,000 year date, which we do not accept, it is no surprise to creationists that early humans were playing music. Genesis 4:21 tells us that Jubal was the father of all who play the lyre and the flute—and he lived not many generations after the first two people had been created.

Sources

The following resources were used in the preparation of this Bible study.

1. *After the Flood*, Bill Cooper
2. *Bones of Contention*, Marvin Lubenow
3. *Catastrophes in Earth History,* Dr. Steven Austin
4. *Creation* magazine (various articles)
5. *Diamonds and the Age of the Earth,* Dr. Vernon Cupps
5. *Dinosaurs by Design,* Duane Gish
6. *Earth Science for Christians Schools Teachers Edition,* Bob Jones University
7. *Frozen in Time*, Michael Oard
8. *Grand Canyon: Monument to Catastrophe,* Dr. Steven Austin
9. *Life in the Great Ice Age,* Michael Oard
10. *NIV Study Bible*, Zondervan
12. *The Answers Book* and *The New Answers Book,* Ken Ham
13. *The Beginning of the World,* Dr. Henry Morris
14. *The Genesis Flood,* Morris and Whitcomb
15. *The Genesis Record*, Dr. Henry Morris
16. *The Young Earth,* Dr. John Morris
17. *Thousands, Not Billions*, the R.A.T.E. Project Team at Institute for Creation Research
18. *Vines Complete Expository Dictionary of Old and New Testament Words*
19. www.answersingenesis.org (various articles)
20. www.icr.org (various Acts & Facts articles)
22. www.probe.org ("Capital Punishment "article)